the
facebook
effect

The Inside Story of the Company
That Is Connecting the World

David Kirkpatrick

Simon & Schuster
New York London Toronto Sydney

Simon & Schuster
1230 Avenue of the Americas
New York, NY 10020

First Simon & Schuster hardcover edition June 2010

SIMON & SCHUSTER and colophon are registered trademarks
of Simon & Schuster, Inc.

For information about special discounts for bulk purchases,
please contact Simon & Schuster Special Sales at
1-866-506-1949 or business@simonandschuster.com.

The Simon & Schuster Speakers Bureau can bring authors to your live event. For more
information or to book an event contact the Simon & Schuster Speakers Bureau at
1-866-248-3049 or visit our website at www.simonspeakers.com.

Designed by Nancy Singer

Manufactured in the United States of America

10 9 8 7 6 5 4 3 2 1

Library of Congress Cataloging-in-Publication Data

Kirkpatrick, David, date.
   The Facebook effect : the inside story of the company that is connecting the world /
David Kirkpatrick.
      p.   cm.
   Includes bibliographical references and index.
   1. Facebook (Firm) 2. Internet industry—United States. 3. Online social networks—
History. I. Title.
HD9696.8.U64F335 2010
338.7'61006754—dc22   2009051983
ISBN 978-1-4391-0211-4
ISBN 978-1-4391-0980-9 (ebook)

To Elena and Clara

Contents

the
facebook
effect

Oscar Morales was fed up. It was holiday time in his hometown of Barranquilla, Colombia, just after the 2008 new year. The gentle-spirited civil engineer with a gift for computers was spending his days at the bucolic nearby beaches with his extended family. But despite the holidays, like much of the country his thoughts were dark, and occupied with the suffering of a little boy named Emmanuel.

Emmanuel was the four-year-old son of Clara Rojas, who had been a hostage in the jungles of Colombia for six years. Her son had been born while she was held by the guerrillas of the Revolutionary Armed Forces of Colombia, known by its Spanish initials, FARC. FARC held a total of seven hundred hostages, including Colombian presidential candidate Ingrid Betancourt, kidnapped along with Rojas during the 2002 campaign.

Sympathy and sadness about the plight of FARC's hostages was an ever-present fact in contemporary Colombia, as was fear about what the powerful and murderous revolutionary army might do next to disrupt the country. But the case of Emmanuel had lately acquired outsized prominence in the popular press. For some time President Hugo Chavez of neighboring Venezuela had been attempting to negotiate with FARC about releasing Betancourt and others. Then abruptly in late December the guerrillas announced that they would soon turn over Rojas, her son Emmanuel, and another hostage to Chavez. In a nation exhausted from a decades-long battle with the violent guerrillas, this was a rare piece of good news. "People were longing for a gift, for a miracle," says Morales, thirty-two. "And Emmanuel was a symbol. The whole country was feeling the promise: 'Please let Emmanuel get his freedom. We would like that as a Christmas present from FARC.'"

But as the new year arrived, Emmanuel still hadn't been freed. Then, in the first days of January, Colombian president Alvaro Uribe went on national television to deliver the shocking news that it appeared that Emmanuel was not even in the possession of FARC! It turned out Emmanuel had become seriously ill some time earlier, and FARC had taken him away from his mother, Clara, and dumped him with a peasant family. He was now, unexpectedly, in the government's hands.

The nation was still on holiday with plenty of time to watch the news, which was all about poor, sick, abandoned Emmanuel. Morales's politically engaged extended family, hanging out by day at the beach, debated what might happen next. "People were happy because the kid was safe, but we were so fucking angry," Morales says. "Forgive me for using that word but we felt assaulted by FARC. How could they dare negotiate for the life of a kid they didn't even have? People felt this was too much. How much longer was FARC going to play with us and lie to us?"

Morales wanted desperately to do something. So he turned to Facebook. Though the service wasn't yet even translated into Spanish, Morales spoke fluent English, as do many educated Colombians, and had been maintaining a profile there for over a year, posting his own information in Spanish and connecting with old college and high school friends. Spending time on Facebook was already a daily ritual for him.

In Facebook's search box he typed the four letters "FARC" and hit enter. There were no results. No groups. No activism. No outrage. Groups devoted to almost everything under the sun were common on Facebook. But when it came to FARC, the citizens of Colombia had become used to being angry but cowed. In effect, the entire country had been taken hostage, and this had been going on for decades.

Morales spent a day asking himself if he was willing to go public on Facebook. He decided to take the plunge, and on the 4th created a group against FARC. "It was like a therapy," he says. "I had to express my anger." He wrote a short description of the group's simple purpose—to stand up against FARC. A self-confessed "computer addict," Morales was skilled at graphics tools, so he designed a logo in the form of a vertical version of the Colombian flag. He overlaid it with four simple pleas in capitals running down the page, each one slightly larger than the last—NO MORE KIDNAPPINGS, NO MORE LIES, NO MORE KILLINGS, NO MORE FARC. "I was trying to scream like if I was in a crowd," he explains. "The time had come to fight FARC. What had happened was unbearable."

But what should he call his group? On Facebook it's conventional to give groups names like "I bet I can find one million people who hate George Bush." But Morales didn't like such titles. They were juvenile. This was not a contest. This was serious. Yet he liked the idea of a million. A famous Spanish song is called "One Million Friends." One million people against FARC? The word *voices* sounded more literary. One million voices against FARC—Un Millon de Voces Contra Las FARC. That was it.

After midnight on January 4, Morales created the group. He made it public so that any Facebook member could join. His personal network included about one hundred friends, and he invited them all. He was tired. At 3 A.M. he went to bed.

At 9 A.M. the next morning he checked his group. Fifteen hundred people had joined already! "Woooooooo!!!" Morales howled in delight. This was an even better response than he had expected! That day at the beach he told his extended family about the group and asked them to invite their own Facebook friends to join. Most of them were avid Facebook users as well, and they hated FARC, too. By the time Morales returned home in the late afternoon, his group had four thousand members.

"That's when I said to myself, 'Okay, no more beach, no more going out.'" He was ready to get serious. "I felt, 'Oh my God! This is what I want! A committed community around the message.'"

A Facebook group has a "wall," where members can post thoughts, as well as discussion forums that allow organized, long-lasting conversations among many members. Morales soon bonded with several people who were posting there with special vigor. They exchanged instant messaging and Skype addresses and cell-phone numbers so they could continue their conversations offline.

As more and more Colombians joined the group, members started talking not only about how mad they were about FARC, but what they ought to do about it. On January 6, just the second full day, a consensus on the page was emerging that the burgeoning group should go public. By the time it hit eight thousand members, people were posting on the discussion board, over and over, "Let's DO something."

Late on the afternoon of the 6th, his newfound Facebook friends, especially two he was speaking to by phone, convinced Morales that he should propose a demonstration. When he did, the idea was received on the wall and discussion board by acclamation. By the end of the day the group, still operating only out of Morales's upstairs bedroom, had decided to stage a national march against FARC. It would be February 4, one month after the formation of the group. Morales, who was used to being left out of things since he lived in a provincial city, insisted the march take place not only in Bogota, the capital, but also many other places throughout the country, including of course his hometown of Barranquilla.

So Morales created an event called the National March against FARC. He and his co-organizers, several of them already as consumed by the project as he was, immediately got pushback from unexpected quarters. Members in Miami, Buenos Aires, Madrid, Los Angeles, Paris, and elsewhere argued that it should be a global demonstration. Morales didn't even realize people living outside Colombia had joined the group. These Colombian émigrés were on Facebook partly to stay in touch with things back at home. They wanted to be involved in this movement, too. So it became a global march.

What ensued was one of the most extraordinary examples of digitally fueled activism the world has ever seen. On February 4, about 10 million people marched against FARC in hundreds of cities in Colombia according to Colombian press estimates. As many as 2 million more marched in cities around the world. The movement that began with an impassioned midnight Facebook post in one frustrated young man's bedroom led to one of the largest demonstrations ever, anywhere in the world.

Facebook's very newness helped Morales's demonstration garner attention in Colombia. Though several hundred thousand Colombians were already using Facebook, it had not appeared on the radar of the average citizen. So when the press began covering plans for the upcoming demonstration, its stories focused heavily on the astonishing impact of this strange American import and the "Facebook kids," as many articles and TV and radio programs called them. Though Morales and his co-organizers were mostly in their early thirties, the country was also capti-vated by the possibility that younger people were not cowed by FARC.

Once Colombian president Alvaro Uribe and Colombia's political establishment saw this Facebook uprising emerge, they did everything they could to make it a success. After a week or two the local army commander began providing Morales with three bodyguards and a car, which he used through February 4. Mayors and city governments throughout the country worked closely with demonstration volunteers to grant march permits.

But what remains remarkable is the way so many Colombians on Facebook signed up on the group under their real names. By the day of the march there were 350,000. Despite decades of fear and intimida-tion, Facebook gave Colombia's young people an easy, digital way to feel comfort in numbers to declare their disgust.

Even after news about the march had become a daily drumbeat in the press and the website had turned into a key promotional tool, Face-book remained central. "Facebook was our headquarters," says Morales. "It was the newspaper. It was the central command. It was the labora-tory—everything. Facebook was all that, right up until the last day."

Morales himself had volunteered to coordinate the local demon-stration in Barranquilla. He expected about 50,000 people to show up. In fact 300,000 did, about 15 percent of the city's population. They filled more than ten city blocks. At exactly noon, Morales read a state-ment that the group had jointly agreed upon. It was broadcast on televi-sion all over Latin America. Demonstrators gathered even in remote places like Dubai, Sydney, and Tokyo. On local TV news, one woman was interviewed in the crush of the Bogota march. Had she been per-sonally injured by FARC, the interviewer asked? "Yes, because I am Colombian," she replied. Morales and his group members had tapped into frustrations deep in the collective national psyche.

While pressure from President Uribe has played a major role in weakening FARC, the demonstrations seem to have struck their own blow. In a sign that the guerrillas were acutely aware of the impending march, on the Saturday before it took place they announced that they would release three hostages, all former Colombian congressmen, as a "humanitarian" gesture. Ingrid Betancourt and fourteen other hostages were rescued in a commando operation by the Colombian army in July 2008. In interviews she recalled listening to a radio in the jungle on February 4, surrounded by her FARC captors. She said she was deeply moved when she heard the demonstrators chanting in unison, "No more FARC! Freedom! Freedom!" Then the guerrillas couldn't stand it and turned off the radio. Oscar Morales is telling me about this in a coffee shop in Manhattan in late 2008. As he does, his voice catches. Tears well up. His group and the subsequent demonstration made him into a national and international celebrity. But the conviction and concern that fueled his creation of Un Millon de Voces Contra Las FARC remains alive. Today he devotes his entire life to the anti-FARC crusade.

Though Facebook was not designed as a political tool, its creators observed early on that it had peculiar potential. During the first few weeks after it was created at Harvard University in 2004, students began broadcasting their political opinions by replacing their profile picture with a block of text that included a political statement. "People were using it back then to protest whatever was important," says Facebook co-founder Dustin Moskovitz. "Even if they were just upset about a minor issue with the school." People from the beginning intuitively realized that if this service was intended as a way for them to reflect online their genuine identity, then an element of that identity was their views and passions about the issues of the day.

"The Colombia thing," says Mark Zuckerberg, Facebook's founder, "is a very early indicator that governance is changing—[and of how] powerful political organizations can form. These things can really affect peoples' liberties and freedom, which is kind of the point of government. . . . In fifteen years maybe there will be things like what happened in Colombia almost every day."

Now, two years after Morales's stunning success, one can find Facebook-fueled activism and protest in every country and community where the service has caught on—and that is pretty much all of them in the developed world. Facebook, along with Twitter, famously played a role in the revolt against the outcome of the mid-2009 elections in Iran. As *New York Times* foreign affairs columnist Tom Friedman pointed out, "For the first time, the moderates, who were always stranded between authoritarian regimes that had all the powers of the state and Islamists who had all the powers of the mosque, now have their own place to come together and project power: the network." It was on Facebook that defeated Iranian presidential candidate Mir Hossein Mousavi told his followers when he thought it was time for them to go into the streets. And when a young woman was tragically killed during one of those protests, it was on Facebook that video of her murder emerged, to be shared worldwide as a symbol of Iranian government repression. The Iranian government, embarrassed, tried several times to shut off access to Facebook. But it is used so widely in the country that it was difficult to do so.

How could Colombia's anti-FARC movement go from A to Z—from one man in his bedroom to millions in the streets—so quickly? Why should Facebook turn out to be a uniquely effective tool for political organizing? How did founder Zuckerberg's decisions at crucial moments in the company's history increase its impact? And in what ways do its unprecedented qualities help explain the rapidity with which Facebook has become a routine part of the lives of hundreds of millions of people around the world? As the rest of this book will explore, many of the answers lie in a set of phenomena I call the Facebook Effect.

As a fundamentally new form of communication, Facebook leads to fundamentally new interpersonal and social effects. The Facebook Effect happens when the service puts people in touch with each other, often unexpectedly, about a common experience, interest, problem, or cause. This can happen at a small or large scale—from a group of two or three friends or a family, to millions, as in Colombia. Facebook's software makes information viral. Ideas on Facebook have the ability to rush through groups and make many people aware of something almost simultaneously, spreading from one person to another and on to many with unique ease—like a virus, or meme. You can send messages to other people even if you're not explicitly trying to. It's how Un Millon de Voces Contra Las FARC grew so fast from its very first night.

Any member who joined was merely making a statement about himself—"Yes, I am against FARC." A new member was not necessarily saying "send this information to my friends," he was just joining the group. But as each new person joined, Facebook took that information and distributed it to the News Feeds of that person's friends. Then when those people joined the group, Facebook reported that news to *their* friends' News Feeds. Something like Morales's anti-FARC campaign that taps into a latent need or desire can spread virally with lightning speed, making groups huge overnight.

Large-scale broadcast of information was formerly the province of electronic media—radio and television. But the Facebook Effect—in cases like Colombia or Iran—means ordinary individuals are initiating the broadcast. You don't have to know anything special or have any particular skills. Twitter is another service with a more limited range of functions that can also enable powerful broadcasting over the Internet by any individual. It too has had significant political impact.

This all may be either a constructive or a destructive force. Facebook is giving individuals in societies across the world more power relative to social institutions, and that may well lead to very disruptive changes. In some societies it may destabilize institutions many of us would rather stay the same. But it also holds the promise—as is starting to be shown in Egypt, Indonesia, and elsewhere—of posing challenges to long-standing repressive state institutions and practices. Facebook makes it easier for people to organize themselves.

There's no reason why the self-organizing component of the Facebook Effect only need apply to serious gatherings, of course. In mid-2008 a Facebook group organized a huge water fight in downtown Leeds, England. And in September 2008 more than a thousand people spent twenty minutes or so smashing each other with pillows in Grand Rapids, Michigan. They heard about the pillow fight on Facebook. Public pillow fights became something of a fad around the world as Facebook-empowered young people embraced a new way to blow off steam.

The Facebook Effect can be no less powerful as a tool for marketers, provided they can figure out how to invoke it, a topic we will explore in greater depth later. Similarly, the Facebook Effect has potentially profound implications for media. On Facebook, everyone can be an editor, a content creator, a producer, and a distributor. All the classic old-media hats are being worn by everyone. The Facebook Effect can create a sudden convergence of interest among people in a news story, a song, or a YouTube video. One day recently I had been working on this book and hadn't paid any attention to the news. I happened to see that a friend's News Feed read "Dow up 3.5%." I would in the past have received that information from Yahoo News, or from radio or television.

The games business, one that is playing a big role in Facebook's development, has already figured a lot of this out. The best games take advantage of the Facebook Effect, with the result that some games are played by as many as 30 million members per week. PlayStation, Xbox, and Nintendo Wii were the platform choices of the previous generation. Now, however, all the video-gaming consoles are starting to build in Facebook connectivity as well.

As Facebook grows and grows past 500 million members, one has to ask if there may not be a macro version of the Facebook Effect. Could it become a factor in helping bring together a world filled with political and religious strife and in the midst of environmental and economic breakdown? A communications system that includes people of all countries, all races, all religions, could not be a bad thing, could it?

There is no more fervent believer in Facebook's potential to help bring the world together than Peter Thiel. Thiel is a master contrarian who has made billions in his hedge fund betting correctly on the direction of oil, currencies, and stocks. He is also an entrepreneur, the co-founder, and former CEO of the PayPal online payments service (which he sold to eBay). He was the very first professional investor to put money in Facebook, in the late summer of 2004, and he's been on the company's board of directors ever since.

"The most important investment theme for the first half of the twenty-first century will be the question of how globalization happens," Thiel told me. "If globalization *doesn't* happen, then there is no future for the world. The way it doesn't happen is that you have escalating conflicts and wars, and given where technology is today, it blows up the world. There's no way to invest in a world where globalization fails." This is a bracing thought, coming as it does from one of the world's great investors. "The question then becomes what are the best investments that are geared towards good globalization. Facebook is perhaps the purest expression of that I can think of."

I had only been marginally aware of Facebook until a public relations person called me in the late summer of 2006 and asked if I would meet with Mark Zuckerberg. I knew that would be interesting, so I agreed. As *Fortune* magazine's main tech writer in New York, I routinely met with leaders of all kinds of technology companies. But when this young man—then just twenty-two—joined me at the fancy Il Gattopardo Italian restaurant in midtown Manhattan, it was at first hard to accept that he was CEO of a tech company of growing importance. He wore jeans and a T-shirt with a line drawing of a little bird on a tree. He seemed so unbelievably young! Then he opened his mouth. "We're a utility," he said in serious tones, using serious language. "We're trying to increase the efficiency through which people can understand their world. We're not trying to maximize the time spent on our site. We're trying to help people have a good experience and get the maximum amount out of that time." He showed no inclination to joke around. He was very focused on commanding my attention for his company and his vision. And he succeeded.

The more I listened the more he sounded like one of the successful—and much older—CEOs and entrepreneurs I talked to regularly in my job. So I casually told him I thought he seemed like a natural CEO. In my mind it was a huge compliment, one I did not give lightly. But he acted insulted. His face scrunched up with a look of distaste. "I never wanted to run a company," he said a few minutes later. "To me a business is a good vehicle for getting stuff done." Then for the rest of the interview he continued to say the kinds of things that only focused and visionary business leaders are capable of saying. From that moment I was confident Facebook's importance would grow. I wrote a column after the meeting called "Why Facebook Matters." The following year my coverage of the company in *Fortune* deepened when Zuckerberg invited me inside the company to do an exclusive story about its groundbreaking transformation into a platform for software applications created by outside entities. That announcement began to change how the world perceived Facebook. By the end of 2007 I had begun to believe Facebook would become one of the world's most important companies. If that was the case, shouldn't somebody write a book about it?

Now a 1,400-person corporation based in Palo Alto, California, Facebook has revenues that could reach $1 billion in 2010. Zuckerberg, now twenty-six, remains CEO. As a result of his determination, strategic savvy, and a fair dollop of luck, he maintains absolute financial control of the company. If he didn't, Facebook would almost certainly today be a subsidiary of some giant media or Internet company. Buyers have repeatedly offered astounding sums of money—billions—if he would agree to sell it. But Zuckerberg is more focused on "getting stuff done" and convincing more people to use his service than he is on getting rich from it. In keeping the company independent he has kept it imbued with his own ideals, personality, and values.

From its dorm-room days, Facebook has looked simple, clean, and uncluttered. Zuckerberg has long had an interest in elegant interface design. On his own Facebook profile he lists his interests: "openness, breaking things, revolutions, information flow, minimalism, making things, eliminating desire for all that really doesn't matter." Despite the founder's interest in minimalism, however, there is much about Facebook that inclines toward excess. Facebook is all information all the time. Each month about 20 billion pieces of content are posted there by members—including Web links, news stories, photos, etc. It's by far the largest photo-sharing site on the Internet, for instance, with about 3 billion photos added each month. Not to mention the innumerable trivial announcements, weighty pronouncements, political provocations, birthday greetings, flirtations, invitations, insults, wisecracks, bad jokes, deep thoughts, and of course, pokes. There's still a lot of stuff on Facebook that probably really doesn't matter.

Popular though it may be, Facebook was never intended as a substitute for face-to-face communication. Though many people do not use it this way, it has always been explicitly conceived and engineered by Zuckerberg and colleagues as a tool to enhance your relationships with the people you know in the flesh—your real-world friends, acquaintances, classmates, or co-workers. As this book explains in detail, this is a core difference between Facebook and other similar services— and has introduced a particular set of challenges for the company at every turn.

The Facebook Effect most often is felt in the quotidian realm, at an intimate level among a small group. It can make communication more efficient, cultivate familiarity, and enhance intimacy. Several of your friends learn from your status update, for example, that you'll be at the mall later. You don't send that information to them. Facebook's software does. They say they'll meet you there, and they show up.

When Facebook is used as it was originally designed—to build better pathways for sharing between people who already know each other in the real world—it can have a potent emotional power. It is a new sort of communications tool based on real relationships between individuals, and it enables fundamentally new sorts of interactions. This can lead to pleasure or pain, but it undeniably affects the tenor of the lives of Facebook's users. "Facebook is the first platform for people," says Esther Dyson, the technology pundit, author, and investor.

Several other factors make Facebook unlike any Internet business that preceded it. First, it is both in principle and in practice based on real identity. On Facebook it is as important today to be your real self as it was when the service launched at Harvard in February 2004. Anonymity, role-playing, pseudonyms, and handles have always been routine on the Web—AOL screen name, anyone? But they have little role here. If you invent a persona or too greatly enhance the way you present yourself, you will get little benefit from Facebook. Unless you interact with others as yourself, your friends will either not recognize you or will not befriend you. A critical way other people on Facebook know you are who you say you are is by examining your list of friends. These friends, in effect, validate your identity. To get this circular validation process started you have to use your real name.

Closely connected to its commitment to genuine identity is an infrastructure intended to protect privacy and give the user control. It doesn't always work, but Zuckerberg and other company officials say they care about it a lot. "Having the friend infrastructure and an identity base ultimately is the key to safety," says Chris Kelly, Facebook's long-time head of privacy, who recently took a leave to run for California state attorney general. "Trust on the Internet depends on having identity fixed and known." If you have doubts about who you are communicating with online, your privacy is at risk. But if you know who is around you, you can authoritatively determine who you would and would not like to see your information.

Privacy, an issue we look at in greater detail in a later chapter, has been a major concern of Facebook's users from the beginning. They often have not felt that it was sufficiently protected, and have periodically revolted in order to say so. Facebook has generally weathered these controversies well. But the issue is fraught—it is a central concern not only of Facebook's users, but, as we will see, of Zuckerberg as well. He knows that Facebook's long-term success will probably be defined by how well it protects its users' privacy. Recently the company has set about simplifying and improving the controls that determine who sees what about you.

The social changes that will be brought about by the Facebook Effect will not all be positive. What does it mean that we are increasingly living our lives in public? Are we turning into a nation—and a world—of exhibitionists? Many see Facebook as merely a celebration of the minutiae of our lives. Such people view it as a platform for narcissism rather than a tool for communication. Others ask how it might affect an individual's ability to grow and change if their actions and even their thoughts are constantly scrutinized by their friends. Could it lead to greater conformity? Are young people who spend their days on Facebook losing their ability to recognize and experience change and excitement in the real world? Are we relying too heavily on our friends for information? Does Facebook merely contribute to information overload? Could we thus become less informed?

What does being a "friend" on Facebook really mean? The average Facebook user has about 130. Can you really have 500 friends, as many do? (I have 1,028, but then I've been writing a book about the company.) What about 5,000, Facebook's maximum? For some, Facebook may generate a false sense of companionship and over time increase a feeling of aloneness. So far there is little data to show how widespread this problem may be, though as our use of electronic media continues in coming years it will certainly remain a widespread concern.

Once I was sitting with Zuckerberg in a modest French bistro a mile or two from Facebook's headquarters, just before closing time. He told me he'd never eaten steak frites before, so I'd urged him to order it. As other tables emptied out, we moved on to coffee and the staff started mopping the floor. Zuckerberg was, as always, wearing a T-shirt, but since it was a little chilly he had on another of his staples—a fleece jacket. I asked him what he thought he was doing when he created Thefacebook (the company's original name) and how his thinking about it had evolved since the early days. His answer was all about transparency. Appropriately enough, Zuckerberg himself is almost compulsively candid.

"I mean, picture yourself—you're in college," he began. "You spend all your time studying theories, right? And you think about things in this abstract way. Very idealistic. Very liberal at this institution. So a lot of these values are just around you: the world should be governed by people. A lot of that stuff has really shaped me. And this is a lot of what Facebook is pushing for.

"Dustin [Moskovitz] and Chris [Hughes] [his Harvard roommates] and I would sit around and talk with other people I was taking computer science classes with. And we'd talk about how we thought that the added transparency in the world, all the added access to information and sharing [enabled by the Internet] would inevitably change big-world things. But we had no idea we would play a part in it. . . . We were just a group of college kids." Then he describes what happened once Thefacebook launched: "Little by little—'Oh, more schools want this'; and 'Okay, more types of people want this.' . . . And it just kept getting wider and wider, and we just went, 'Wow.'

"Then one day it kind of hit us that we could play a leading role in making this happen and pushing it forward. . . . And what seemed obvious to my group of friends who were just armchair intellectuals talking about this in college—about how transparency coming from people would transform how the world works and how institutions were governed—it was like, 'Hey, maybe other people *aren't* actually pushing this, and maybe it takes this group of people who grew up thinking these things and having these values to push it forward. Maybe we shouldn't give up.'" And he laughs.

Mark Zuckerberg was never one to defer to authority figures. Facebook started out as his own revolt against Harvard's unwillingness to build an online facebook. But what he built turns individuals into the authority. The entire service revolves around the profile and the actions of people. Facebook empowers them at the expense of institutions. In building it, Zuckerberg transferred a little bit of his own power to all the service's members.

Facebook is bringing the world together. It has become an overarching common cultural experience for people worldwide, especially young people. Despite its modest beginnings as the college project of a nineteen-year-old, it has become a technological powerhouse with unprecedented influence across modern life, both public and private. Its membership spans generations, geographies, languages, and class. It may in fact be the fastest-growing company of any type in history. Facebook is even bigger in countries like Chile and Norway than it is in the United States. It changes how people communicate and interact, how marketers sell products, how governments reach out to citizens, even how companies operate. It is altering the character of political activism, and in some countries it is starting to affect the processes of democracy itself. This is no longer just a plaything for college students.

If you use the Internet, you are increasingly likely to use Facebook. It is the second-most-visited site, after Google, and claims more than 400 million active users (as of February 2010). Well over 20 percent of the 1.7 billion people on the Internet worldwide now use Facebook regularly. Facebook added high school students in fall 2005 and opened to everyone in fall 2006. Now users around the world spend about 8 billion minutes there every day (the average user spends almost an hour each day on Facebook). And even despite all its growth, the number of people there is growing at a mind-bending rate—about 5 percent a *month*. Were the growth rates of both Facebook and the Internet to remain steady, by 2013 every single person online worldwide would be on Facebook.

Of course that will never happen. But Facebook already operates in seventy-five languages, and about 75 percent of its active users are outside the United States. About 108 million Americans are active on Facebook, or 35.3 percent of the entire population, according to the *Facebook Global Monitor,* published by InsideFacebook.com. That sounds impressive. But 42 *percent* of Canada's population uses it. The largest number of Facebook users is still in the United States, but the next ten countries are a global mix. In order, they are the United Kingdom, Turkey, Indonesia, France, Canada, Italy, the Philippines, Spain, Australia, and Colombia. The ten countries in which it grew fastest in the year ending February 2010, according to the *Facebook Global Monitor,* were Taiwan, the Philippines, Vietnam, Indonesia, Portugal, Thailand, Brazil, Romania, Lithuania, and the Czech Republic.

Unlike just about any other website or technology business, Facebook is profoundly, centrally, about people. It is a platform for people to get more out of their lives. It is a new form of communication, just as was instant messaging, email, the telephone, and the telegraph. During the early days of the World Wide Web, people sometimes said that everyone would eventually have their own home page. Now it's happening, but as part of a social network. Facebook connects those pages to one another in ways that enable us to do entirely new things.

But this scale, rate of growth, and social penetration raise complicated social, political, regulatory, and policy questions. How will

Facebook alter users' real-world interactions? How will repressive governments respond to this new form of citizen empowerment? Should a service this large be regulated? How do we feel about an entirely new form of communication used by hundreds of millions of people that is completely controlled by one company? Are we risking our freedom by entrusting so much information about our identity to one commercial entity? Tensions around these questions will grow if Facebook keeps extending its influence across more and more of the globe.

This book aims to explore these questions. But you can only understand how Facebook became such an amazing company and where it might go if you understand how it all got started in a dormitory in Cambridge, Massachusetts, as the brainchild of a restless and irreverent nineteen-year-old kid.

1

The Beginning   "We have opened up Thefacebook for popular consumption at Harvard University."

Sophomore Mark Zuckerberg arrived at his dorm room in Harvard's Kirkland House in September 2003 dragging an eight-foot-long whiteboard, the geek's consummate brainstorming tool. It was big and unwieldy, like some of the ideas he would diagram there. There was only one wall of the four-person suite long enough to hold it—the one in the hallway on the way to the bedrooms. Zuckerberg, a computer science major, began scribbling away.

The wall became a tangle of formulas and symbols sprouting multicolored lines that wove this way and that. Zuckerberg would stand in the hall staring at it all, marker in hand, squeezing against the wall if someone needed to get by. Sometimes he would back into a bedroom doorway to get a better look. "He really loved that whiteboard," recalls Dustin Moskovitz, one of Zuckerberg's three suite-mates. "He always wanted to draw out his ideas, even when that didn't necessarily make them clearer." Lots of his ideas were for new services on the Internet. He spent endless hours writing software code, regardless of how much noncomputing classwork he might have. Sleep was never a priority. If he wasn't at the whiteboard he was hunched over the PC at his desk in the common room, hypnotized by the screen. Beside it was a jumble of bottles and wadded-up food wrappers he hadn't bothered to toss.

Right away that first week, Zuckerberg cobbled together Internet software he called Course Match, an innocent enough project. He did it just for fun. The idea was to help students pick classes based on who else was taking them. You could click on a course to see who was signed up, or click on a person to see the courses he or she was taking. If a cute girl sat next to you in Topology, you could look up next semester's

Differential Geometry course to see if she had enrolled in that as well, or you could just look under her name for the courses she had enrolled in. As Zuckerberg said later, with a bit of pride at his own prescience, "you could link to people through things." Hundreds of students immediately began using Course Match. The status-conscious students of Harvard felt very differently about a class depending on who was in it. Zuckerberg had written a program they wanted to use.

Mark Zuckerberg was a short, slender, intense introvert with curly brown hair whose fresh freckled face made him look closer to fifteen than the nineteen he was. His uniform was baggy jeans, rubber sandals—even in winter—and a T-shirt that usually had some sort of clever picture or phrase. One he was partial to during this period portrayed a little monkey and read "Code Monkey." He could be quiet around strangers, but that was deceiving. When he did speak, he was wry. His tendency was to say nothing until others fully had their say. He stared. He would stare at you while you were talking, and stay absolutely silent. If you said something stimulating, he'd finally fire up his own ideas and the words would come cascading out. But if you went on too long or said something obvious, he would start looking through you. When you finished, he'd quietly mutter "yeah," then change the subject or turn away. Zuckerberg is a highly deliberate thinker and rational to the extreme. His handwriting is well ordered, meticulous, and tiny, and he sometimes uses it to fill notebooks with lengthy deliberations.

Girls were drawn to his mischievous smile. He was seldom without a girlfriend. They liked his confidence, his humor, and his irreverence. He typically wore a contented expression on his face that seemed to say "I know what I'm doing." Zuck, as he was known, had an air about him that everything would turn out fine, no matter what he did. It certainly had so far.

On his application for admission to Harvard two years earlier, he could barely fit all the honors and awards he'd won in high school—prizes in math, astronomy, physics, and classical languages. It also noted he was captain and most valuable player on the fencing team and could read and write French, Hebrew, Latin, and ancient Greek. (His accent was awful, so he preferred ancient languages he didn't have to speak, he told people with typical dry humor.) Harvard's rarefied social status was neither intimidating nor unfamiliar. He'd attended the elite Phillips Exeter Academy, where you are expected to proceed to the Ivy League. He'd transferred there as a kind of lark. He'd gotten bored after two years at a public high school in Dobbs Ferry, New York, north of New York City.

Zuckerberg is the second-oldest of four children of a dentist father and a psychologist mother, and the only boy. The family home, though the largest in the neighborhood, remains modest. Its dental office in the basement is dominated by a giant aquarium. The elder Zuckerberg, something of a ham, is known as "painless Dr Z." His website announces "We cater to cowards," and a sign outside the home office shows a satirical scene of a wary dental patient. Mark's sisters, like him, are academic stars. (His older sister Randi is now a senior marketer at Facebook.) From his early years Zuckerberg had a technical bent: the theme of his bar mitzvah was "Star Wars."

The suite was one of the smallest in Kirkland House. Each of the two bedrooms came with bunk beds and a small desk. Zuckerberg's roommate was Chris Hughes, a handsome, tow-headed, openly gay literature and history major with an interest in public policy. They dismantled the bunks—it was fairer, they decided, if nobody had to sleep on top. But now the two single beds took up almost all the space. There was hardly room to move. The desk was useless anyway—it was piled high with junk. In the other bedroom was Moskovitz, a hardworking, Brillo-haired economics major who was himself no intellectual slouch, and his roommate, Billy Olson, an amateur thespian with an impish streak.

Each boy had a desk in the common room. In between were a couple of easy chairs. It was, like the entire suite, a mess. Zuckerberg had a habit of accumulating detritus on his desk and nearby tables. He'd finish a beer or a Red Bull, put it down, and there it would stay for weeks. Occasionally Moskovitz's girlfriend would get fed up and throw out some garbage. Once, when Zuckerberg's mother visited, she looked around the room embarrassedly and apologized to Moskovitz for her son's untidiness. "When he was growing up he had a nanny," she explained.

This warren of tiny rooms on the third floor pushed the boys toward greater intimacy than they might have shared under less constrained conditions. Zuckerberg was by nature blunt, even sometimes brutally honest—a trait he may have acquired from his mother. Though he could be taciturn he was also the leader, simply because he so often started things. A habit of straight talk became the norm in this suite. There weren't a lot of secrets here. The four got along in part because they knew where each stood. Rather than getting on one another's nerves, they got into one another's projects.

The Internet was a perennial theme. Moskovitz, who had little training in computing but a natural penchant for it, kept up a constant repartee with Zuckerberg about what did and did not make sense on-line, what would or would not make a good website, and what might or might not happen as the Internet continued its inroads into every sphere of modern life. At the beginning of the semester, Hughes had zero interest in computing. But by midyear he too had become fascinated by the constant discussion of programming and the Internet, and started chiming in with his own ideas, as did Moskovitz's roommate, Olson. As Zuckerberg came up with each new programming project, the other three boys had plenty of opinions on how he should build it.

In the common room of Suite H33 in Kirkland House, Ivy League privilege and high geekdom converged. What happened there turns out not to have been common, but at the time it seemed pretty routine. Zuckerberg was hardly the only entrepreneur beavering away on a business in his dorm room. That wasn't too noteworthy at Harvard. Down every hall were gifted and privileged children of the powerful.

It's presumed at Harvard that these kids are the ones who will go on to rule the world. Zuckerberg, Moskovitz, and Hughes were just three eggheads who loved to talk about ideas. They didn't think much about ruling the world. But from their funky, crowded dorm room would emerge an idea with the power to change it.

Emboldened by the unexpected success of Course Match, Zucker-berg decided to try out some other ideas. His next project, in October, he called Facemash. It gave the Harvard community its first look at his rebellious irreverent side. Its purpose: figure out who was the hottest person on campus. Using the kind of computer code otherwise used to rank chess players (perhaps it could also have been used for fencers), he invited users to compare two different faces of the same sex and say which one was hotter. As your rating got hotter, your picture would be compared to hotter and hotter people.

A journal he kept at the time, which for some reason he posted along with the software, suggests Zuckerberg got onto this jag while upset about a girl. "——— is a bitch. I need to think of something to make to take my mind off her," he wrote, adding "I'm a little intoxi-cated, not gonna lie." Perhaps that pique is what led him to the idea, mused about in the journal, of comparing students to farm animals. Instead, according to the journal, Billy Olson came up with the idea of comparing people to other people and only occasionally putting in a farm animal. By the time the program launched, the animals were gone completely. "Another Beck's is in order," Zuckerberg wrote as he continued his Facemash chronicles. The entire project was completed in an eight-hour stretch that ended at 4 A.M., said the journal.

The photos for the Facemash website came from the so-called "facebooks" maintained by each of the Harvard houses where under-graduates live. They were the pictures taken the day students arrived for orientation—the kind of clumsy, awkwardly posed shots almost every-one would prefer to disavow. Zuckerberg cleverly found ways to obtain digital versions from nine of Harvard's twelve houses. Student newspa-per the *Harvard Crimson* later called it "guerrilla computing." In most cases he was able to simply hack in over the Web. At Lowell House a friend gave Zuckerberg temporary use of his log-in. (The friend later re-gretted it.) At another house, Zuckerberg snuck in, plugged an Ethernet cable into the wall, and downloaded names and photos from the house computer network.

The fact that he was doing something slightly illicit gave Zucker-berg little pause. He could be a touch headstrong and liked to stir things up. He didn't ask permission before proceeding. It's not that he sets out to break the rules; he just doesn't pay much attention to them.

He started running the Facemash website on his Internet-connected laptop in mid-afternoon of Sunday, November 2. "Were we let in [to Harvard] for our looks?" the site asked on its home page. "No. Will we be judged by them? Yes." Zuckerberg emailed links to a few friends, later claiming he had only intended them to test it out and make suggestions. But once people started using it, they apparently couldn't stop. His "testers" alerted their own friends and Facemash became an instant underground hit.

The *Crimson* somewhat eloquently opined on the appeal of the software afterward, even as its editorial scolded Zuckerberg for "catering to the worst side of Harvard students": "A peculiarly-squinting senior and that hottie from your Medieval manuscripts section—click! Your blockmate and the kid who always glared at you in Annenberg—click! Your two best friends' respective significant others—pause . . . click, click, click! . . . We Harvard students could indulge our fondness for judging those around us on superficial criteria without ever having to face any of the judged in person." Yes, it was fun.

One gay resident of a suite near Zuckerberg's was elated when, in the first hour, his photo was rated most attractive among men. He of course alerted all his own friends, who then started using the site. When Zuckerberg returned to his room at 10 P.M. from a meeting, his laptop was so bogged down with Facemash users that it was freezing up. But neighbors were not the only ones suddenly paying attention to Facemash. Complaints of sexism and racism quickly started circulating among members of two women's groups—Fuerza Latina and the Association of Harvard Black Women. Quickly the computer services department got involved and turned off Zuckerberg's Web access. By the time that happened, around 10:30 P.M., the site had been visited by 450 students, who had voted on 22,000 pairs of photos.

Zuckerberg was later called before Harvard's disciplinary Administrative Board, along with the student who'd given him the password at Lowell House, his suite-mate Billy Olson (who, as the online journal noted, had contributed ideas), and Joe Green, a junior who lived in the

next suite through the fire door, who had helped out as well. Zuckerberg was accused of violations of the college's code of conduct in the
way the site handled security, copyright, and privacy. The board put
him on probation and required him to see a counselor, but decided
not to punish the others. If Zuckerberg hadn't omitted the farm animal
photos, he probably wouldn't have gotten off so lightly. He apologized
to the women's groups, claiming he had mainly thought of the project
as a computer science experiment and had no idea it might spread so
quickly.

Green's father, a college professor, happened to be visiting his son
the night Zuckerberg was celebrating his comparatively light sentence
for Facemash. The sophomore had gone out and bought a bottle of
Dom Perignon, which he was exultantly sharing with his Kirkland
neighbors. Says Green: "My dad was trying to drill it into Mark's head
that this was a really big deal, that he'd almost gotten suspended. But
Mark didn't want to hear it. My dad came away with the notion that I
shouldn't do any more Zuckerberg projects." It would later prove to be
a very expensive prohibition.

But to everyone else, the episode was a clear sign: Zuckerberg had
a knack for making software people couldn't stop using. That came as
little surprise to his roommates. They knew he had even been talking
to Microsoft and other companies about selling a program he'd written with a friend as his senior project at Exeter, called Synapse. The
software watched what kind of music someone liked so it could suggest other songs. His friends called the program "The Brain" and were
especially excited when they heard Zuckerberg might get as much as a
million dollars for it. If that happened, they pleaded, could he please
buy a large flat-screen TV for the common room?

Zuckerberg kept making little Web programs, like one he created
quickly to help himself cram for his Art in the Time of Augustus course.
He had barely attended the class all first semester. As the final loomed,
he cobbled together a set of screens with art images from the class. He
emailed the other class members an invitation to log in and use this

study aid and add comments alongside each image. His classmates took his cue. After they all used it, he spent an evening scrutinizing what they'd said about the images. He passed the final. He also wrote a program he called "Six Degrees of Harry Lewis," an homage to a favorite computer science professor. He used articles in the *Harvard Crimson* to try to identify relationships between people, and created a whimsical network of connections to Lewis based on these links. You could type in any Harvard student's name and the software would tell you how they were connected to Professor Lewis.

He also worked on other people's projects. After the Facemash episode he mended fences with the Association of Harvard Black Women by helping them set up their own website. And he worked for a while with three seniors who aimed to build a dating and socializing site they called Harvard Connection. They had an idea for a service that would tell you about parties and provide discounted admission to nightclubs, among other intended features. But they weren't programmers. The three, athletic six-foot-five-inch identical twins Cameron and Tyler Winkelvoss, both champion rowers on the crew team, and their friend Divya Narendra sought out Zuckerberg in November after reading about Facemash in the *Crimson*. They offered to pay him to do the programming for their service.

"I had this hobby of just building these little projects," says Zuckerberg now. "I had like twelve projects that year. Of course I wasn't fully committed to any one of them." Most of them, he says, were about "seeing how people were connected through mutual references."

Zuckerberg's interest in building websites with social components had arisen the previous summer. He was living in a dormitory at the Harvard Business School with two Exeter friends, including Adam D'Angelo, with whom he'd developed Synapse, the music suggestion software, and who was now studying computer science at the California Institute of Technology. Another close friend and Harvard computer science major named Kang-Xing Jin lived there, too. All three had lucrative programming jobs they found undemanding, and Zuckerberg had broken up with his girlfriend. There was a lot of time for bull sessions, which tended to center on what kind of software should happen next on the Internet.

D'Angelo had launched a provocative project of his own the previous year from his dorm room at Caltech. Called Buddy Zoo, it invited users to upload their AOL Instant Messenger (AIM) buddy lists to a server and compare them to the lists belonging to people who had also uploaded theirs. You could see who shared which friends, thus illustrating your network of social connections. At the time AIM was the de facto communications tool of American youth (and many adults). Hundreds of thousands of AIM users tried out Buddy Zoo, and it had a brief online celebrity. D'Angelo made no effort to commercialize it, and eventually let it die. But it pointed in a promising direction.

During the winter break, Zuckerberg got deep into coding yet another project. He was particularly eager to get this one done. His bemused friends didn't pay much more attention to this new project than to all the other sites he had launched that year.

On January 11, Zuckerberg went online and paid Register.com thirty-five dollars to register the web address Thefacebook.com for one year. This site borrowed ideas from Course Match and Facemash as well as from a service called Friendster that Zuckerberg belonged to. Friendster was a social network, a service that invited individuals to create a "profile" of themselves, complete with data about hobbies, tastes in music, and other personal information. On such services people linked their own profiles to those of their friends, thereby identifying their own "social network."

Friendster, like most social networks up to that time, was primarily intended to help you connect with people for dating. The idea was that you might find romantic material by scrutinizing the friends of your friends. Friendster had taken Harvard by storm the previous year but had fallen from favor after its almost overnight nationwide success led to millions of users. That created technical strains that made it slow and difficult to use. Another, more flashy social network called MySpace had launched the previous August in Los Angeles. It was growing quickly and already had about a million members, though it hadn't made much of an impression at Harvard.

Harvard had been claiming for many months that it was going to take all the "facebooks" maintained by each House—the ones Zuckerberg had cannibalized for Facemash—and unify them online in searchable form. Studying these photos was a common recreational activity. There was a college-wide printed facebook called the Freshman Register, issued each year, but it only included entering students. Copies were extensively annotated—boys, for instance, would circle photos of the best-looking girls.

Now that students had seen what was possible with Friendster, they wanted an online facebook. It was obvious that it wasn't that hard to create online directories. If an entrepreneur in San Francisco could do it, why couldn't Harvard's administration? This impulse was surprisingly widespread. At many colleges that year, students were pushing administrations to put student photo directories online. The *Crimson* included extensive references to the need to create an online facebook. The editors took the view that if a student could build Facemash, there was no reason a programmer couldn't build a facebook. In a December 11 editorial titled "Put Online a Happy Face: Electronic facebook for the entire College should be both helpful and entertaining for all," its editors practically described how to build one. The essay strongly emphasized the need for students to control their own information in such a system. That fall Zuckerberg took a math class on graph theory. At semester's end everyone in the class went out to dinner and ended up talking about the need for a "universal facebook." So Zuckerberg went home and built one.

"There was definitely a little bit of a 'fuck you' to Harvard," says one classmate and friend of Zuckerberg's. "They always said they were going to do a centralized facebook, but they had all these worries about how it's their information. They thought they had legal issues. Mark just figured you could get people to upload the information themselves." In fact, Zuckerberg later said that it was the *Crimson*'s editorials about Facemash that gave him the initial idea for how to build Thefacebook. "Much of the trouble surrounding the facemash could have been eliminated," wrote the *Crimson*, "if only the site had limited itself to students who voluntarily uploaded their own photos."

That simple insight, combined with Zuckerberg's desire to create a reliable directory based on real information about students, became the core concept of Thefacebook. "Our project just started off as a way to help people share more at Harvard," says Zuckerberg, "so people could see more of what's going on at school. I wanted to make it so I could get access to information about anyone, and anyone could share anything that they wanted to."

His new service for Harvard students was not a dating site like Friendster. It was a very basic communications tool, aimed at solving the simple problem of keeping track of your schoolmates and what was going on with them. Some of Zuckerberg's friends later speculated that it was also intended to help him deal with his own introverted personality. If you're a geek who is a little uncomfortable relating to other people, why not create a website that makes it easier?

Thefacebook also drew inspiration from another important source—the so-called away messages that users of AIM posted when they weren't at their computers. These short, pithy phrases were often used by AIM users to show off their creativity. Though there was room for only a few words, users included political statements and humor as well as practical information about the account holder's whereabouts. AIM away messages were so important to Zuckerberg that another one of his earlier software projects was a tool that alerted him when friends' messages changed. Thefacebook was going to be a robust combination of the AIM away message and that alert tool—a place where you could host more information about yourself so friends could keep track of you. (Today's Facebook status update traces its heritage directly back to those AIM away messages.)

Both Course Match and Facemash had operated over Zuckerberg's dorm-room Net connection from his laptop, but Course Match's success had taken its toll on the hard drive. Zuckerberg lost quite a lot of data. And part of what got him in trouble with the administrative board over Facemash was that he used Harvard's network to host it. So this time he took a more serious approach. He searched around online and found a hosting company called Manage.com, where he entered his credit card number and started paying eighty-five dollars a month for space on a computer server. That's where Thefacebook's software and data would reside. This would be Thefacebook.com, not part of the www.harvard.edu network. He wasn't sure, but in the back of his mind, Zuckerberg had a notion that this could end up as more than just a brief entertainment.

Here's another sign he thought something unusual might happen: he made a deal with a business-savvy classmate, Eduardo Saverin, to give him one-third of Thefacebook in exchange for Saverin making a small investment and helping out with business matters. Zuckerberg knew Saverin from Alpha Epsilon Pi, a selective fraternity for Jewish students to which both had recently pledged. Saverin was supposed to figure out how Thefacebook, if it took off, could make some money. The polished and well-liked son of a wealthy Brazilian business magnate, Saverin was an officer in the college Investment Club and a superb chess player who was known by his friends as a math genius. The two nineteen-year-olds agreed to invest $1,000 each. (Joe Green says Zuckerberg also approached him to be a business partner, but when Professor Green heard about it, he got "kind of pissed," so Joe declined. Later he took to calling it, always with a pained laugh, his "billion-dollar mistake.")

On the afternoon of Wednesday, February 4, 2004, Zuckerberg clicked a link on his account with Manage.com. Thefacebook.com went live. Its home screen read: "Thefacebook is an online directory that connects people through social networks at colleges. We have opened up Thefacebook for popular consumption at Harvard University. You can use Thefacebook to: Search for people at your school; Find out who are in your classes; Look up your friends' friends; See a visualization of your social network." Zuckerberg labeled himself user number four. (The first three accounts were for testing.) User number five was roommate Hughes; number six was Moskovitz; and number seven was Saverin. Zuckerberg's friend and classmate Andrew McCollum designed a logo using an image of Al Pacino he'd found online that he covered with a fog of ones and zeroes—the elementary components of digital media.

The software spread quickly from the very beginning. The first users—Zuckerberg's Kirkland House neighbors—sent emails to other students asking them to join and become their friends. That begat other emails from those students inviting their own friends to join. Someone suggested sending an email to everyone on the Kirkland House mailing list—about three hundred people. Several dozen signed up almost immediately.

Thus began a viral explosion. By Sunday—four days after launch—more than 650 students had registered. Three hundred more joined on Monday. Thefacebook almost instantly became a main topic of conversation in Harvard dining halls and between classes. People couldn't stop using it.

To sign up, you created a profile with a single picture of yourself, along with a bit of personal information. You could indicate your relationship status. Pick one from the drop-down menu: single, in a relationship, or in an open relationship. You could include your phone number, AIM username, and email address; indicate which courses you were taking (a feature inspired by Course Match); favorite books, movies, and music; clubs you belonged to; political affiliation: very liberal/liberal/moderate/conservative/very conservative/apathetic; and a favorite quote. Thefacebook had no content of its own. It was merely a piece of software—a platform for content created by its users.

Privacy controls were part of the original design. And there were some big restrictions: you couldn't join unless you had a Harvard.edu email address, and you had to use your real name. That made Thefacebook exclusive, but it also ensured that users were who they said they were. Zuckerberg later told the *Crimson* that he "hoped the privacy options would help to restore his reputation following student outrage over facemash.com." Validating people's identity in this way made Thefacebook fundamentally different from just about everything else that had come before on the Internet, including Friendster and MySpace. On Thefacebook you could set your privacy options to determine exactly who could see your information. You could limit it to current students, just people in your class, or only those in your residential house.

Once you set up your own profile, the interaction began. It was pretty limited. After you invited others to be your friend, you could see a diagram of your social network, which showed all the people you were connected to. You could also direct something called a "poke" at other users by simply clicking on a link on their page. When you did, an indication would show up on their home page. What did that mean? Here's the insouciant answer Zuckerberg posted on the site: "We thought it would be fun to make a feature that has no specific purpose. . . . So mess around with it, because you're not getting an explanation from us."

Much activity on Thefacebook from the beginning was driven by the hormones of young adults. It asked you whether you were "interested in" men or women. In addition to giving you the option to list whether you were in a relationship, you were asked to fill in a section labeled "Looking for." One frequently chosen option was "Random play." When you poked someone, an indication of that simply showed up on their profile. That person could poke you back. For at least some, the interaction had a distinctly sexual meaning. This was college, after all.

Many people, on the other hand, found practical and wholesome uses for Thefacebook—creating study groups for classes, arranging meetings for clubs, and posting notices about parties. Thefacebook was a tool for self-expression, and even at this primordial stage of its development people were starting to recognize that there were many facets of the self that could be projected on its screen.

Another feature was timely for many students. You could click on a course and see who was taking it, just as in Course Match. At Thefacebook's launch, students were in the middle of choosing courses for the following semester. It was what's called "shopping week" at Harvard, when classes have begun but students can add or drop them at will. For any Harvard student who picked his or her courses partly based on who else was in class, this feature of Thefacebook was immediately useful. It helps explain the rapid spread of Thefacebook in its early days, and also why Zuckerberg launched it that week.

The whiteboard by the bedrooms in Kirkland Suite H33 now took on a different, less abstract character. Zuckerberg began filling it with charts and graphs indicating Thefacebook's growth—how many people were joining each day and what features they used. It also tracked which users had the most friends.

On Monday the 9th the *Crimson* interviewed Zuckerberg, something its staff was becoming accustomed to. "The nature of the site," he told the paper, "is that each user's experience improves if they can get their friends to join it." Still smarting from the rebuke he received for Facemash, he emphasized to the *Crimson* that he was "careful . . . to make sure that people don't upload copyrighted material." The *Crimson* did a little probing about his motives: "Zuckerberg . . . said he did not create the Website with the intention of generating revenue. . . . 'I'm not going to sell anybody's e-mail address,' he said. 'At one point I thought about making the Website so that you could upload a resume too, and for a fee companies could search for Harvard job applicants. But I don't want to touch that. It would make everything more serious and less fun.' "

Making Thefacebook fun was more important than making it a business. It was a statement that would reverberate down through the short history of Facebook.

Thefacebook may have been meant to replace the Harvard house facebooks, but from the beginning there was one obvious difference. Whereas photos taken by college photographers the first week of school were often awkwardly posed, poorly lit, and unflattering, the ones people posted of themselves on Thefacebook tended to cast them in a very positive light. These were the young superstars of tomorrow, as envisioned by themselves. In only the second article ever written about Thefacebook, on February 17 a prescient columnist for the *Crimson* pinpointed several characteristics that would forever after form a central part of Facebook's appeal. Wrote junior Amelia Lester (who five years later would be named managing editor of the *New Yorker*): "While Thefacebook.com isn't explicitly about bringing people together in romantic unions, there are plenty of other primal instincts evident at work here: an element of wanting to belong, a dash of vanity and more than a little voyeurism."

And competitiveness was immediately in evidence. From Theface-book's first day, some users thought of it more as a way to accumulate the largest possible number of friends than to communicate and gather useful information. Many users of Facebook still do.

By the end of the first week, about half of all Harvard undergraduates had signed up, and by the end of February approximately three-fourths. But students were not the only ones showing their faces online. The only requirement for membership was that you have a Harvard email address, which meant Thefacebook was available not only to students—graduate as well as undergrad—but also to Harvard alumni and staff. Some students griped that the staff didn't belong there. While only a few had joined so far, about a thousand alumni had, mostly recent ones. After three weeks Thefacebook had more than 6,000 users.

Within days, Zuckerberg realized that he was going to need help to operate and maintain Thefacebook. So he turned to those closest at hand—his roommates. About a week after Thefacebook launched, Zuckerberg signed an employment contract with Dustin Moskovitz. A year later, in a talk, he told the story of Moskovitz joining this way: "One of my roommates was 'Hey, I'll help you!' I said 'Dude! You can't pro-gram!' So he went home for the weekend and bought the book PERL for Dummies and said 'Now I'm ready.' I said 'Dude, the site's not writ-ten in PERL.'" Regardless, Zuckerberg adjusted the ownership of The-facebook to give the eager Moskovitz 5 percent. He reduced his own stake slightly to 65 percent and Saverin's to 30 percent. Moskovitz's main job was to spearhead expansion to other campuses.

From even the second week, students at schools other than Harvard were emailing Zuckerberg asking when they could have it, too. Moving beyond Harvard had been in Zuckerberg's mind from the beginning. Even the home page implied it—"an online directory that connects people through social networks at colleges"—not "Harvard," but "col-leges." And his ambition did not stop there. Moskovitz says that while he was hired to help add new schools, "in that same conversation it was like—'Yeah, and then we'll go beyond that.'"

Moskovitz mimicked Zuckerberg's code wherever he could, and set out to learn. He wasn't always fast, but he immediately became known for his amazing capacity for hard work. "Mark would get kind of impatient," says one friend. "But Dustin just trudged through and through and through." Some in Kirkland House started calling the sophomore from Florida "the ox."

Zuckerberg now says Moskovitz's role during this period was "critical" to Thefacebook's success. To add a school, Moskovitz had to figure out how email was addressed for students, staff, and alumni so he could set up the registration procedure. Then he would obtain a list of courses and residential dorms. He also had to set up a link to the college newspaper, because Thefacebook then had a feature, later discontinued, that linked your profile to any article in the campus paper that mentioned you. It took about half a day to do all the legwork and coding to add each school, but Zuckerberg and Moskovitz started expanding to new ones quickly even though both were still taking a full course load. They opened to students at Columbia on February 25, to Stanford the next day, and to Yale on the 29th. Columbia started slowly, but Stanford is where the broad appeal of Thefacebook was first proven. After just a week, the *Stanford Daily* was writing that "Thefacebook.com craze has swept through campus." It reported 2,981 Stanford students had already registered.

Zuckerberg hated doing interviews and talking in public, but he gave the *Stanford Daily* a lot of time. "I know it sounds corny, but I'd love to improve people's lives, especially socially," he told the paper. He also said that since the site was still only costing him eighty-five dollars per month, he didn't feel any business imperative: "In the future we may sell ads to get the money back, but since providing the service is so cheap, we may choose to not do that for a while."

He didn't want to do many interviews like that in the future. The newspaper at every new school seemed to want to talk to him, and the guys were planning to add a lot of colleges. So shortly thereafter, Zuckerberg recruited yet another likely prospect, his own roommate, Chris Hughes. Hughes became Thefacebook's official spokesman. The company's founding quartet was complete. Thefacebook had 10,000 active users. It had been operational for one month.

f

As Thefacebook grew at Harvard, Zuckerberg continued to disavow any serious business motivation. But once he began extending it to other schools, he started showing strategic instincts that would befit a CEO, as well as a steely willingness to confront competition. The reason he decided to expand first to Columbia, Stanford, and Yale, he says now, is that those three schools each already had its own homegrown social networks. It was a sort of market test—putting his product up against the best competition that was out there. "If TheFacebook still took off at those schools and displaced those [other networks] then I would know it would go really well at all the other ones," he explains.

At Stanford, Thefacebook took off like a rocket. A school-specific social network there called Club Nexus had already mostly flamed out. When students saw Thefacebook, it felt to many like exactly what they'd been waiting for. "It wasn't something that had to be explained," says a 2005 graduate.

But at Columbia, a student named Adam Goldberg had launched a commercial site called CUCommunity a month before Zuckerberg created Thefacebook. By the time Thefacebook came to Columbia four weeks later, 1,900 of the school's 6,700 undergraduates were active on CUCommunity. It would be several months before Thefacebook overtook it. CUCommunity also soon started expanding to other schools. At Yale, the student-run College Council had launched a dating website and online facebook called YaleStation on February 12. Though it had fewer features than Thefacebook, it was experiencing a similar stratospheric uptake—by the end of the month about two-thirds of undergraduates had registered.

But Zuckerberg was convinced that his service had legs, so he decided to extend it further into the Ivy League—launching at both Dartmouth and Cornell on Sunday, March 7. At Dartmouth, a friend of Zuckerberg's from Exeter was chair of the Student Assembly's Student Services Committee. Like student governments at Harvard, Penn, Yale, and other schools, it had been lobbying to put the campus facebook online. The friend agreed to promote Thefacebook using the assembly's email system to all students. That message went out at 10 P.M. By the following evening, 1,700 of Dartmouth's 4,000 undergraduates were users.

The speed of adoption got Zuckerberg so excited he agreed again to talk to the college newspaper, the *Dartmouth*. "It blows my mind that people have actually used the site," he told the paper. "I'm all about people expressing, and however people see fit to use the site, that's cool." Zuckerberg had also gotten help at Stanford, where a childhood friend from Dobbs Ferry provided him with a password to get into the Stanford network as well as a list of student email addresses and dormitories.

Pretty quickly, though, it was more about fending off interest than stimulating it. Emails started to arrive from around the country, begging Zuckerberg and crew to bring Thefacebook to other schools. Within weeks the four Harvard sophomores—all still carrying a full course load—had launched their service at MIT, University of Pennsylvania, Princeton, Brown, and Boston University. By mid-March the total user number hit 20,000. Yet another high school classmate of Zuckerberg's at Exeter entered the picture. This time it was Adam D'Angelo, Exeter's other programming whiz, Zuckerberg's summer roommate, and co-author of music recommendation program Synapse. From his dorm at Caltech, D'Angelo helped Moskovitz do programming to add new schools. The Ivy League and similar schools were the first to launch largely because that's where the real-world social networks of users at Harvard could be found—mostly friends from high school. Thefacebook had an elite edge.

Up until now, it had been designed so that within each school, users could see one another's profiles unless they chose not to. You could deliberately ratchet up your privacy settings, but most students didn't. Any user from Harvard, for example, could see most Harvard students' profiles. That was the default. Harvard students could not, however, see profiles of students at Stanford. But for Thefacebook to continue to grow, it would need cross-campus linking, and there was a growing chorus of complaints that it wasn't possible. So Zuckerberg and Moskovitz decided that such links could be created by the mutual agreement of both people. This became the template for how Facebook connections are established to this day.

As costs mounted, Zuckerberg mused to the *Crimson*, which had taken to idolizing him, that "it might be nice in the future to get some ads going." By the end of March, with the active-user number surpassing 30,000, Thefacebook was paying $450 per month for five servers from Manage.com. Zuckerberg and Saverin each agreed to invest another $10,000 into the company. Meanwhile, Saverin had begun selling a little advertising and had secured a few small contracts with companies that sold moving services, T-shirts, and other products to college students. These ads began to appear in April.

It was harder and harder simply to keep Thefacebook operating smoothly. Thousands of users could be online at once, straining the servers. Zuckerberg and Moskovitz tried to delay adding new schools until they had worked out kinks for the users they already had. "Our growth to other colleges was always constrained by our server capacity," recalls Moskovitz. "We simply could not scale the architecture fast enough." Luckily they could hold off on new schools until they resolved the kinks. The two programmers were continually re-architecting how the site operated and working to make it more efficient. Moskovitz was trying to pick up as much as he could from the more experienced Zuckerberg, and from D'Angelo, twenty-five hundred miles away at Caltech.

Zuckerberg now marvels with gratitude when he recalls Moskovitz's dedication in those days. "Dustin took the competition so seriously," he says. "I'd be like 'Hey, I heard through the grapevine that this other service is thinking of launching at this college.' He'd be 'Really? No way!' And that paper he was supposed to be doing he'd just like scrap it and go and launch at that school. He was just a workaholic and a machine. Early on I viewed it as a project. I wasn't super-invested in it because it wasn't clear to me it was going to be this huge thing. I was like 'Yeah, this is pretty neat. It's not the end-all be-all, but it's cool. And I have these other classes.' But Dustin joined and really helped scale it."

The boys were using free open-source software like the MySQL database and Apache Web server tools, which made the entire undertaking affordable. But while the software might have been free, it was not simple to operate. Zuckerberg was a more practiced programmer than Moskovitz, but he had never operated these kinds of programs before. He was learning by the day, even as he studied for four courses, including a demanding one in computer science. But so popular was

Thefacebook that by the end of the semester, each time they added a new school its students signed up almost en masse.

Zuckerberg had a burning desire to try new things, but his ability to create a fast-growing website in his spare time had a lot to do with where he was situated. "Having genius and ambition alone isn't going to get you there. It's really important to be lucky," says Moskovitz. "But Mark had all three in spades, including luck. He just fell into the right situations a lot, and had extremely good timing. And when he saw a good idea he wanted to pursue it, whereas another person might have felt he needed to finish school first."

Facebook's ultimate success owes a lot to the fact that it began at college. That's where people's social networks are densest and where they generally socialize more vigorously than at any other time in their lives. Moskovitz actually studied this question during that fateful spring semester. He wrote a paper for a statistics class using data from Thefacebook. It demonstrated that, as he describes it, "any individual student is within two degrees of everybody else on a given campus." On average, students are separated from each other by no more than one intermediary relationship. "That's why Thefacebook grew so well in college," explains Moskovitz. He got an A in the statistics class, which wasn't bad considering that the majority of his time that semester was spent working on the site. "And I got a bunch of sweet-ass bonus points for the data set," Moskovitz remembers with relish.

Harvard offered Zuckerberg unique resources for developing his business. "At Harvard people were starting up websites pretty frequently," says Moskovitz. "Even an impressive hedge fund—people were doing that as undergrads. So it wasn't that crazy to say 'my roommate happens to like to do these big consumer websites.'" Several others, like the Winkelvoss/Narendra team, were even working on social networks.

And the sheer talent that existed among Zuckerberg's roommates was extraordinary. There aren't many schools where he could find someone as talented as Moskovitz in the bedroom on the other side of the wall. The two hadn't met until move-in day at the beginning of the year, but Zuckerberg found in his suite-mate not only a hard-working programmer but an intellectual and leader who would effectively serve as Facebook's chief technology officer for years. Likewise, Chris Hughes, his own roommate, was so articulate and polished that he served as Facebook's spokesman. Later Hughes played an important role in the 2008 presidential campaign of Barack Obama.

Then of course there's the allure of something that began in the most exclusive halls of all academia. Harvard confers an imprimatur that carries unique weight in any field. A Harvard connection makes a product less suspect. To join a social network that began at Harvard might seem perfectly natural to anyone with a high opinion of himself. That was an important early dynamic.

It also didn't hurt that Harvard students are preternaturally status-conscious. The service served as a validation of the scale of your social ambitions even as it measured your success. Sam Lessin, a Zuckerberg friend and classmate who was an early user, says, "There is incredible latent social competition at Harvard which I think really helped Facebook in the early days." If people were going to maintain their profile and social networks online, then the kind of natural elitists who attend Harvard had no compunction attempting to construct the best and largest of them. Back in that *Crimson* opinion piece written when Thefacebook was less than two weeks old, Amelia Lester nailed it: "There's little wonder why Harvard students, in particular, find the opportunity to fashion an online persona such a tantalizing prospect. Most of us spent our high school careers building resumes so padded they'd hold their own in a sumo match, an experience which culminated in the college application. . . . Most of all [Thefacebook] is about performing . . . and letting the world know why we're important individuals. In short, it's what Harvard students do best."

But some proffer a darker narrative for how and why Zuckerberg got Thefacebook started at Harvard. By these accounts, Zuckerberg is a thief, and Thefacebook was the idea of other Harvard students. The most serious accusation is one made by Cameron and Tyler Winkelvoss and Divya Narendra. The trio say Zuckerberg stole numerous ideas from their plan for Harvard Connection after they hired him to program it. After a month or two of work, Zuckerberg concluded that their plan was unlikely to succeed. Shortly thereafter he began work on Thefacebook. This disagreement would become an expensive problem for Zuckerberg's nascent company.

In mid-April 2004, over two months after the site had gone live, business manager Saverin, now calling himself the company's chief financial officer, took steps to formalize Thefacebook as a business. He set up a limited-liability company in Florida, where he had attended high school. The partners listed were Zuckerberg, Moskovitz, and Saverin.

Though revenues for Thefacebook were nonexistent in its first weeks, by mid-February Zuckerberg had already begun fielding calls from people interested in investing. They'd heard about the extraordinary growth of this new site and wanted to get a piece. At the end of the semester, classmate Lessin, whose father was a well-known investor, took Zuckerberg around New York to meet with venture capitalists and executives in the finance and media industries.

At one of those meetings in June, a financier offered Zuckerberg $10 million for the company. Mark had just turned twenty. Thefacebook was four months old. He didn't for a minute think seriously about accepting.

2

Palo Alto "Founder, Master, and Commander, Enemy of the State."

As the Spring 2004 Harvard semester wound down, things at Thefacebook just got busier. By the end of May it was operating at thirty-four schools and had almost 100,000 users.

In June 2004, business manager Saverin opened a bank account and deposited more than $10,000 of his own money as working capital and also began depositing advertising revenues there.

A month or so earlier, Saverin had gotten in touch with a firm called Y2M. Y2M sold ads for college newspaper websites and Saverin invited them to come talk about selling ads for Thefacebook. The meeting was delayed a couple of times because Mark and Eduardo had exams or papers due. When Y2M's Tricia Black finally sat down with them, Zuckerberg pulled out a notebook with a printout of Thefacebook's traffic data. Black was nonplussed. "You must be tracking it wrong," she said. "There's no way you could have this much traffic." Zuckerberg suggested that the ad company put their own monitoring software on the server for a few days to track it for themselves.

The stunning numbers weren't an error. Black and her colleagues were thrilled. Y2M began almost immediately placing ads for clients, taking a commission of around 30 percent. One of the first advertisers was MasterCard, seeking applications for a special credit card for college students. But like Y2M itself and most of its other advertisers, MasterCard executives were skeptical Thefacebook could really deliver results. So instead of simply paying to display ads, as it did with this campaign on other college sites, MasterCard agreed to pay only when a student filled out a card application. At this point Thefacebook operated at about twelve schools. MasterCard turned on its campaign at

5 P.M. on a Thursday. Within one day it received twice the applicants it had expected for the entire four-month campaign. Thefacebook was getting ads in front of exactly the right customers—wealthy undergrads at the best schools. MasterCard continued advertising.

Y2M's executives started to see Thefacebook as a potential game-changer and by summer they wanted a piece of the action. Black and another executive met with Zuckerberg and asked if Y2M could invest. The young CEO said he'd consider it, but that they would have to give Thefacebook a valuation of at least $25 million. Y2M decided to hold off.

In situations like this, Zuckerberg tends to be impassive. Often he says little, despite facing extravagant praise or seductive entreaties. He was unimpressed with Y2M's advances. Even then he had his own vision for the potential of Thefacebook, and it didn't have a lot to do with money. "We're going to change the world," Black remembers him saying. "I think we can make the world a more open place." These were words he would speak again and again in coming years.

Maximizing revenue by selling ads was less important to Zuckerberg than keeping users happy. He would allow advertisements, but only on his terms. Advertisers could only use a few standard-size banners. Those who requested customized treatments were refused. Zuckerberg turned down ads from companies he thought were out of keeping with the playful student mood of Thefacebook, including Mercer Management Consulting and Goldman Sachs. Zuckerberg for a while even put little captions above the display ads reading "We don't like these either but they pay the bills." Says Joshua Iverson, a sales rep who worked for Black at Y2M: "Mark never wanted ads. Eduardo was the businessperson." Of course it wasn't unusual for contemporary Web thinkers to be uninterested in advertising. Sites like Craigslist and Wikipedia were at that time rapidly becoming among the Internet's largest by taking a patently noncommercial approach.

Y2M tried to convince Zuckerberg to expand Thefacebook onto campuses with larger student populations, like the University of Arizona. But he was resolute it remain mostly Ivy League, or at least limited to schools his users were asking him to add—places where their friends went to school. This kept the circle small and exclusive those first few months. Advertisers themselves couldn't even log on to Thefacebook.com, since Zuckerberg insisted membership remain limited to students, faculty, and alumni of the schools where it operated. It was unheard-of for advertisers not to be able to see their own ads running.

But despite these challenges, Black was growing ever more certain that Thefacebook was a sure thing. After Y2M itself failed to get a piece of Thefacebook, she started campaigning for Saverin to hire her full-time.

Meanwhile, Zuckerberg was hedging his bets. He didn't take Thefacebook's success for granted. In fact, while he had high hopes, he still wasn't certain the website would amount to much. He still looked at it as just one of his projects, although it was becoming an interesting one. So, ever the entrepreneur, he embarked on yet another new project. While he still spent most of his nonstudy time on Thefacebook, he and Andrew McCollum, another talented sophomore programmer, started working on new software they called Wirehog. Inspired in part by the once-notorious music-sharing site Napster, Wirehog was going to be a peer-to-peer content-sharing service. It would allow users not only to exchange music, but video and text files or any kind of digital information—and only with friends. It would connect directly to Thefacebook, turning your friends there into sources for content.

Zuckerberg searched the Craigslist classifieds and found a four-bedroom ranch house in Palo Alto, California, which he rented as a summer sublet. He decided he wanted to go out to California for several reasons. McCollum, with whom he was collaborating on Wirehog, had a summer internship at nearby video game company Electronic Arts, an industry giant that had created the Sims, Madden NFL games, and many other hits. Exeter buddy Adam D'Angelo was willing to come up from Caltech to hang out. But most of all it was the promised land of technology. "Palo Alto was kind of like this mythical place where all the techs used to come from," he told a reporter a few months later. "So I was like, I want to check that out."

In a critical recruitment effort, Zuckerberg convinced Dustin Moskovitz to join him on the trip to California. Moskovitz had already arranged a summer job in the Harvard computer lab as a user assistant, or UA. But Moskovitz had become indispensable. With his dogged work ethic and growing knowledge of coding, he was more or less managing Thefacebook's day-to-day operation. Zuckerberg promised to pay more than he'd get in his UA job and convinced him the move would be good for Thefacebook.

Spokesman and Zuckerberg roommate Chris Hughes had already paid for a summer program in France and would only come out to Palo Alto when that was over. His middle-class North Carolina family didn't have a lot of money and he was by nature even more risk-averse than Moskovitz, whose Florida family was fairly well-off. The more worldly Brazilian Saverin had his own reasons not to join the trek to Palo Alto, which appealed to him not at all. He headed to New York for the summer, planning to drum up more advertising business and to work at an investment firm where his father had connections.

Sean Parker was stressed out. It was a hot afternoon in Palo Alto, and he hated doing physical work. But his lease was up and he was short on cash. So here he was in June 2004 on the sidewalk in front of his girlfriend's family's house, unloading boxes from his car. It was, admittedly, a svelte vehicle—a white BMW 5-series he'd bought when times were flush. Parker too was a bit svelte. His curly blond hair was stylishly long. The slim twenty-four-year-old wore a fashionable and pricey T-shirt, which on this day was getting sweaty.

When he noticed a group of boys heading toward him he stiffened. His boxes contained expensive computer equipment. He didn't like the look of these kids—all wearing sweatshirts with hoods up despite the heat. He thought they had a menacing air, maybe a group of hoodlums. But now the shortest one walked right up.

"Parker!" he said unexpectedly, with enthusiasm. "Sean—it's Mark, Mark Zuckerberg." Suddenly it all snapped into place. This was the guy he'd met for dinner in New York two months earlier. He'd said he was coming out to California for the summer.

Zuckerberg introduced the other four—all Harvard undergrads, not hoodlums: Thefacebook's curly-haired co-founder Dustin Moskovitz, Andrew McCollum, Zuckerberg's Wirehog partner, and two skinny interns that Thefacebook had hired for the summer, Harvard freshmen Erik Schultink and Stephen Dawson-Haggerty. The five boys had been on a mile-long walk home from the grocery store, since they didn't have a car. They were living in a house just a block away. Zuckerberg invited Parker to come over. A few hours later, the young entrepreneur walked to the Thefacebook house at 819 La Jennifer Way.

Sean Parker was about to become a major—if controversial—character in the Facebook story. He had a lot of Internet experience for someone his age. In 1999 he'd hooked up online with a guy named Shawn Fanning, the creator of Napster, and then joined him in San Francisco to help launch the service that upended the music industry. Parker left Napster after just a year and co-founded his own Internet company, Plaxo. The venture quickly raised millions and began garnering hundreds of thousands of users, but Parker ran into trouble again with his financial backers. Plaxo's venture capitalists didn't like his casual approach to scheduling and deadlines, his iconoclasm, his insecurity, or his superior attitude, though they recognized he was scary smart. The investors didn't much appreciate Parker's rock-and-roll lifestyle, either. He would work weeks on end to accomplish some company objective, sleeping in the office, then not come in at all for days. Finally they booted him out. In the end they even hired a private investigator to document his alleged misbehavior.

Parker was among the growing number of Silicon Valley executives who were becoming convinced that social networking would become a very big business. In the fall of 2003, Silicon Valley venture investors had put a total of $36 million into four high-profile social networking start-ups—Friendster, LinkedIn, Spoke, and Tribe. In late March, not long after Thefacebook took over the Stanford campus in mere days, Parker sent Zuckerberg an email out of the blue. He played up his Napster bona fides and offered to introduce Zuckerberg to savvy San Francisco investors who understood social networking. He mentioned that he was acquainted with the CEOs of LinkedIn and Tribe, who had jointly purchased a key patent that might be important for social networks. Parker suggested that a meeting with them could help ensure that the patent wasn't used against Thefacebook. Saverin emailed him back, and they arranged a dinner in New York.

In early April, Parker flew to New York for the dinner. He joined Zuckerberg, Zuckerberg's girlfriend Priscilla Chan, Saverin, and Saverin's girlfriend at a trendy new Chinese place called 66 in Tribeca. Zuckerberg was thrilled to meet a founder of Napster, which he considered one of the most important things that had ever happened on the Internet. And Parker was quickly impressed with Zuckerberg. At the sleek, Richard Meier–designed restaurant, the two fell into intense back-and-forth almost immediately, mostly leaving out Saverin and the two women. Zuckerberg sketched out his vision for what Thefacebook could become. It was an even bigger vision than Parker had expected. "He was not thinking, 'Let's make some money and get out,'" says Parker. "This wasn't like a get-rich-quick scheme. This was 'Let's build something that has lasting cultural value and try to take over the world.' But he didn't know what that meant. He was a college student. Taking over the world meant taking over college." Parker remembers thinking Zuckerberg seemed incredibly ambitious. "He had imperial tendencies." Parker had to overdraw his bank account to afford the dinner, but he felt it was worth it.

When he ran into Parker two months later on the Palo Alto sidewalk, Zuckerberg had a strong and positive recollection of the New York meeting. Parker was one of the people who seemed to really understand what Thefacebook was doing.

Over dinner in Palo Alto, Zuckerberg witnessed the denouement of Parker's months-long battle with his former backers at Plaxo. The six young men walked to a nearby restaurant, where Zuckerberg brought Parker up to date on Thefacebook and introduced him more fully to his Harvard chums. While they were sitting in the restaurant, Parker got a critical call from his lawyer. The news was bad. The Plaxo board had decided not to allow about half of Parker's remaining Plaxo shares to vest. In other words, he was getting kicked out of his company and losing his chance to make any money if it later went public or was sold.

Parker was enraged. He was getting screwed. Thefacebook's boys listened in awe and dismay. It became the theme for the night. Zuckerberg had little experience dealing with investors, though they had been approaching him regularly since about March, hoping to get a piece of Thefacebook. Hearing Parker's story was chastening. "VCs sound scary," Zuckerberg recalls thinking. It was a formative moment, and a critical one for Facebook's future. Feeling for his friend, and thinking he might learn much from Parker, Zuckerberg invited him to move into the house with them. By September, Zuckerberg was calling Parker the company's president.

Parker is a unique sort of entrepreneur, even for Silicon Valley. A precocious programmer and intellect, he is the son of a top U.S. government oceanographer. He spent much of his Virginia childhood beset by illness, devoting much of his time to reading and learning computer programming. In 1995 he became an intern at fifteen at Freeloader, one of Washington, D.C.'s first Internet start-ups. Several years later, in 1999, barely out of high school, he helped Shawn Fanning start Napster. The renegade peer-to-peer music-sharing service attracted 26 million users by its peak in early 2001. It was also the first big consumer service to demonstrate a fundamentally new sort of Internet—one where users connected directly to one another without a big company like eBay or Yahoo or Microsoft in the middle. But Napster almost immediately encountered an all-out legal assault from the big record labels. Parker, for his part, lost his job there in a management shake-up after little more than a year, when he was still just twenty. He got the company in trouble by openly discussing in emails—displayed in a court case brought by the labels—that what Napster's users were doing might be illegal. Shortly thereafter, he and two friends formed Plaxo, which helped users keep track of email addresses and contact information.

Despite his lack of formal education and loose respect for business norms, Parker is a business intellectual. He could even perhaps be called a business artist, if those two words can be juxtaposed. On his own Facebook profile he calls himself "a twisted half-breed: a rational-aesthete." He combines a subtle understanding of business history, economics, and behavior with an artist's impatience, impulsiveness, and vision for a better world. Not that his actual vision is any good. His eyes are bad enough that if he forgets his contacts or his thick glasses he can need help getting around. He has a certain weightless quality, as if he were about to float off like Peter Pan, perhaps surrounded by one of his always-gorgeous girlfriends. (Lately he has settled into a long-term relationship.)

A voracious reader with a deep fascination with politics, the self-taught Parker may pepper an analysis of current trends with a reference to "the intentions of the framers" (the men who wrote the U.S. Constitution, that is). His Facebook profile includes quotes from T. S. Eliot, Bertrand Russell, and Albert Camus. He likes to talk about things like "business externalities." And if you show the slightest interest he will eagerly describe his theory on the history of media since Gutenberg. Most of all, he likes to talk, rapidly, intensely, and he likes to talk about ideas. What he brought to Thefacebook was both a practiced understanding of the realities of business and a penchant for philosophical argument that prompted Zuckerberg to refine his vision. Hanging with Parker wasn't that different from jawboning with classmates in the Harvard dorms, except that the conversation now was all about making Thefacebook successful.

The boys quickly settled into a routine—sleep late, walk into the dining room, and get to work. The table there was piled high with computers, cables, modems, cameras, and trash that got stuffed among them, along with the requisite untossed bottles, cans, and cups. Zuckerberg slept later than most—he seldom got to work before afternoon, and usually worked well into the night. His typical garb in this office of sorts was pajama bottoms and a T-shirt. When they sat at their laptops around the dining room table on La Jennifer Way, it was eerily quiet. That's because when they did talk to one another, they did it over instant messaging, even when they were sitting right next to one another. It let the others concentrate. Geeks like Zuckerberg and Moskovitz like to get deep into what is almost a trance when they're coding, and while they didn't mind background music or the TV playing, they couldn't stand interruptions.

With Moskovitz and Parker, Zuckerberg had now put into place, consciously or not, an ideal team to bracket his own talents. Moskovitz is the kind of person every start-up needs—diligent, down-to-earth, versatile, and pragmatic. He took responsibility for keeping the service operating and setting up databases for new schools (with the interns doing much of the tedious work). If he had to, he'd work all night to keep the system up.

Parker, by contrast, was an experienced company-maker, familiar with the ways of the world. He specialized in networking in the real-world sense. He knew a lot of people in the Valley and understood how to get their ear. He was polished—spending money (when he had it) on nice meals, haircuts, and stylish clothes. He might occasionally cancel meetings unexpectedly after burning himself out at a party the night before, but he was a slick front man who could talk up Thefacebook, which was exactly what it needed. In Silicon Valley those who had heard of it still mostly thought of Thefacebook as a silly thing for sex-starved college kids. Parker's big-picture vision helped give the service gravitas.

Having the two of them in place meant Zuckerberg could do what he does best—think about what Thefacebook should be and how it should evolve. Or, depending on his mood, devote his energies to something he wanted to use himself—Wirehog. Ironically, Zuckerberg was not a heavy user of Thefacebook. Nor, in fact, were any of its founders and early employees. This summer the interns, working with Moskovitz, started to gather data on how people actually used the site. They found that some users were looking at hundreds and even thousands of profiles every day. These were the users they were designing for.

When he wasn't working on Wirehog, Zuckerberg was coding a feature for Thefacebook he also thought would be pretty cool—a way to get information out of the service using short messages, or SMS, from a cell phone. Way before there was a Facebook application for the iPhone or the BlackBerry, this was Thefacebook's mobile interface. You could send messages with a person's name to m@Thefacebook.com and include special codes to get friends' phone numbers or other information sent back to your phone. The only problem was that it was unwieldy for ordinary users. You needed to carry around a folded-up cheat sheet to remember how to use it. Cool though it was, it didn't last long.

Parker moved into a room with a bare mattress on the floor. Zuckerberg later said that aside from his car, the only impressive thing Parker brought with him was a pair of "ridiculously nice sneakers." According to Parker, here's how Zuckerberg asked him to take on the president job: "Can you help us set up the company? We're screwed right now." Part of the deal, though, was a quid pro quo—Parker got to stay in the house; Zuckerberg and his friends got to share Parker's BMW.

At least one adviser urged Zuckerberg not to hire Parker, saying his lax ways and dissipated lifestyle could taint the company. "He has a problem with women and rock 'n' roll," Zuckerberg's more-experienced confidant said. But Zuckerberg was unmoved. He said he'd heard the stories, but that Parker's experience and intelligence outweighed the risks. After all, Parker helped start Napster. Not only that, but he was a small investor in Friendster and a friend of its founder. He was already talking about The facebook as "the chance to do Friendster correctly."

Thefacebook's traffic had dipped now that college kids were mostly out of school. But Zuckerberg and Moskovitz were bolstering the site for fall, when they expected growth would resume in earnest. Some visitors saw their confidence as arrogance bred of upper-class Harvard privilege. "Even back then they were talking like they knew this was going to happen and they had the best thing in the world and they were going to dominate everyone," says one slightly awed early visitor to the house. "They used that word *dominate* all the time." Thefacebook would dominate its rivals, they said. In fact, much of it was bluster, with an added dash of the insouciance of youth.

Work would get intense in the late afternoon and early evening. "Everyone would be working and someone would say 'Hey, I'm hungry. I wanna go get In-N-Out,'" says another frequent visitor, "and Mark would, like, pound the table and just say 'No! We're in lockdown! No one leaves the table until we're done with this thing.'" Like *dominate*, *lockdown* became a part of Facebook's lingo and lore that lasted for years.

Despite his baby face and general shyness, Zuckerberg was firmly and undisputedly in charge. Every page of Thefacebook included at the bottom a little tagline: "A Mark Zuckerberg production." On the service's "about" page, he was listed as "Founder, Master and Commander, Enemy of the State." Moskovitz, by contrast, had the relatively ignominious listing "No Longer Expendable Programmer, Paid Assassin." Saverin's job was said to be "Business Stuff, Corporate Stuff, Brazilian Affairs."

Zuckerberg was beginning, fitfully, to show qualities of natural leadership. Says Sean Parker: "The leader of a company needs to have a decision tree in his head—if this happens we go this way, but if it winds up like that, then we go this other way. Mark does that instinctively." He liked to have fun as much as any of his colleagues—in fact he could be a bit of a comic—but he also was determined to keep this ship moving forward. And he was more than happy to be the captain.

Not infrequently, in fact, he acted like he was captain of a pirate ship. When he started thinking hard about something or was debating an idea with one of the others, Zuckerberg would often jump up and start pacing back and forth around the room with his hands clasped behind his back. Among the few possessions he had brought out with him were his fencing paraphernalia, which he left lying in a pile not far away. Often he'd grab his foil and start swinging it through the air. "Okay, we've got to talk about this," he would declare, one hand held behind his back, lunging forward with his foil. It got on Moskovitz's nerves. "I'm the personality type where that would get me sometimes," says Moskovitz. "It was a pretty small room. I'm like a cautious mother— 'You're going to break something!' But when he got into the mood he would do it for a couple hours." Later Moskovitz and the others banned fencing from the house.

Behind the house was a nice kidney-shaped pool, and the triangular backyard was mostly paved. One night Zuckerberg and Parker spent a few hours standing around outside talking. Zuckerberg had his foil, and was waving it too close to Parker's face for comfort. He found it very bad for concentration to have a fencing sword swing a few inches from his face every few seconds. "Do you think this thing is really going to last?" Zuckerberg asked at one point between thrusts. "I do," replied Parker, wincing. "Unless we get outcompeted by somebody else or we don't execute well or we let our servers fail like Friendster did, there's no reason why this can't last."

"Mark was actually very rational about the low probability of building a true empire," recounts Parker. "He had these doubts. Was it a fad? Was it going to go away? He liked the idea of Thefacebook, and he was willing to pursue it doggedly, tenaciously, to the end. But like the best empire builders, he was both very determined and very skeptical. It's like [former Intel CEO] Andy Grove says, 'only the paranoid survive.'"

Adam D'Angelo, up from Caltech, was by far the most gifted and accomplished programmer of the bunch, but he was working on his own projects. He was also neither expert in nor very interested in the relatively simple Web-based languages Thefacebook was employing—PHP, JavaScript, and HTML. D'Angelo had a bad case of carpal tunnel syndrome, which meant his hands and arms hurt when he typed. So he was trying to come up with his own alternative—invent a way to move his hands in the air that a video camera could recognize, in order to manipulate text on a screen. It was a pretty challenging project, maybe too challenging, and as summer went on he spent less time on that and more time helping McCollum and Zuckerberg with Wirehog.

While the young engineers worked to bolster the site and refine its features, Parker started thinking about what it would mean to turn Thefacebook into a company. He hired the lawyer who'd helped set up Plaxo. He started looking for someone to manage "operations," a fundamental task in Internet companies that involves making sure the data center and servers are operating properly. Up until then, all that work had been outsourced to third-party companies, but Thefacebook was getting too big for that. Parker discovered that his young colleagues didn't even know the basics about network management, like what a router was. He found an engineer named Taner Halicioglu, who had experience at eBay. He worked from home in San Jose.

Parker became Thefacebook's front man, especially with investors. It wasn't uncommon for fancy cars to be parked outside on the dead-end street, under the big droopy trees that loomed over the front of the house. That meant someone with money was inside. Some guys from the Benchmark venture capital firm wanted to know if there was a chance of an equity investment. The answer was no, for the time being.

But Thefacebook was going to need more funding in the near future, so Parker made sure such people felt comfortable calling or stopping by.

A couple of Google executives came over to see if there might be a way to work with or even buy Thefacebook. Even at this early date, Google was well aware that something noteworthy was going on in Palo Alto. Zuckerberg and Parker were leery, though, because the risk of becoming subsumed by Silicon Valley's Internet giant was real. If they wanted to do their own thing, they had to stay independent, they believed. Anyway, what they were trying to do was very different from what Google did. Their site was about people; Google was about data.

One area where Parker and Zuckerberg clashed was over Wirehog, on which development continued. The new president thought it was a huge distraction from the work of growing Thefacebook. And his history with Napster made him leery of getting into another tussle with music and media companies. To Parker it seemed likely that such companies would accuse Wirehog—and with it Thefacebook—of helping users steal content, just as the music industry had with Napster. With Wirehog engineer McCollum, the two flew down to Los Angeles where they met with Edgar Bronfman, Jr., CEO of Warner Music Group, and Tom Whalley, who ran Warner Bros. Records. Parker had gotten to know Whalley in his Napster days. Unsurprisingly, they were wholeheartedly opposed to Wirehog. Though Parker feared that a successful lawsuit against Wirehog could take Thefacebook down along with it, he failed to sway Zuckerberg, who persevered.

"Really great leadership," says Parker, "especially in a start-up, is about knowing when to say no—evoking a vision very clearly, getting everybody excited about it, but knowing where to draw the line, especially with products. You can't do everything. And that's a lesson Mark didn't know yet. That's a lesson Mark learned."

Work was hardly the only priority, of course. What group of twenty-year-olds suddenly occupying their own house wouldn't want to party? Nerds these guys might have been, but they were fun-loving nerds. Stanford was just a mile or so away. It operates on a quarter system, so students were still around in the summer. Using a feature in Theface-book that enabled ads to be targeted at just one school, the housemates announced their parties right on the service—"Thefacebook is having a party!"—and then often found themselves mobbed by both Stanford students and townies. Moskovitz started dating a girl who had just graduated from Palo Alto High School.

The parties were typical beer-and-booze-fueled affairs. Here's where Parker came in particularly handy. He was the only one in the group over twenty-one, so they relied on him to buy the alcohol. There was a fair amount of pot smoking, too, though Zuckerberg frowned on it and didn't partake. "Mark is just about the most anti-drug person I've ever met," says one friend.

Hanging out around the pool was of course a major activity. If a glass broke, the shards often just got swept into the water. McCollum strung a wire from the chimney on top of the house to a spot slightly lower on a telephone pole beyond the pool. With a pulley, he turned it into a zip line, so you could ride down the wire and, suspended over the pool, drop in with a massive splash.

One favorite party activity was Beirut, or beer pong, a beer-drinking game for teams of two or more players that involves throwing a Ping-Pong ball into a bunch of beer cups arrayed in a triangle at the other end of a table. If you get your ball into the opposing team's cup, they have to drink that cup's contents. Once all the losing team's cups are eliminated, its members drink the remaining beer on the winning team's side. The losers get really drunk.

Beirut was so popular at Thefacebook (and at Harvard) that six months later Zuckerberg and friends launched a national college Beirut tournament. Thefacebook planned to pit campus teams against one another, then each school's winning team was to come to New York for the final to compete for a $10,000 prize. (The *Stanford Daily* asked Zuckerberg why Thefacebook would host an event it had to spend $10,000 on, and he replied "Because it's cool.") Thousands of students paid ten dollars each to register, but Thefacebook canceled the competition only four days after it launched, after being deluged with complaints from colleges.

The house felt like a dorm. They'd often grill hamburgers or steaks out by the pool, and eat, raucously, at an outdoor table. If the talking lasted too late at night, the neighbors would get peeved. If someone brought a girl back to his room, his roommate had to sleep downstairs on the couch or drag his mattress into another room. Some people—female and male—stopped by for days and just hung around.

One of them was a friend of Parker's named Aaron Sittig. He had earlier helped create a version of Napster for the Macintosh, called Macster, which Napster bought. At this point he was working for a nascent music-oriented social network called Imeem, located a few blocks away in Palo Alto. Sittig is a quiet, self-effacing, blond surfer type who in addition to being a programmer is a superb graphic designer and typographer. But back then he was feeling burned-out and unmotivated. Parker brought him around because he thought he could help Thefacebook, especially with design.

But Sittig wasn't showing a lot of initiative. "I kept explaining to Mark that Aaron was brilliant," says Parker. "But Aaron would just sit on the couch and diddle around on his computer all day playing with fonts. Mark kept saying 'Who is this guy? He's worthless. He doesn't do anything.' Mark thought it was bad for the work ethic to have him hanging around seeming to do nothing." (The following year, after reenrolling at the University of California at Berkeley for a semester to study philosophy, Sittig did come to work at Thefacebook. He became one of Zuckerberg's closest confidants.)

Oftentimes the coding, swordplay, and raucous meetings would go on well into the night, sometimes punctuated by breaks for drinking, movie-watching, and video-game playing. The Xbox got a workout, with the game Halo a particular favorite. Somehow Tom Cruise became a group obsession, and thus ensued a lengthy Tom Cruise movie marathon. They rented an entire stack of his DVDs. Why Tom Cruise? Sittig, who put down his laptop long enough to watch along with everyone else, explains: "Tom Cruise was funny because he's not a very cool character. He's not a cool guy." It was camp.

Pretty soon they were naming the servers on which Thefacebook's software was running after characters in Tom Cruise movies: "'Where's that script running?' 'It's running on Maverick.' 'Well, run it instead on

Iceman, I need Maverick to test this feature.'" (Maverick and Iceman were characters in Cruise's 1986 film *Top Gun.*) The Ben Stiller movie *Zoolander* was another house favorite, watched to excess. It played over and over in the background while people were working. These guys found it funny to quote big chunks of the movie to one another. They may have been developing a revolutionary Internet service, but they were still really just college kids.

With a total of seven guys living in the house, they needed more than Parker's BMW to get around, so Zuckerberg and company bought a car. They were planning to return to Harvard in the fall so expected to sell it again in three months. They spent a few hundred dollars on one they thought couldn't depreciate further—a forest-green, twelve-year-old, manual-transmission Ford Explorer. It was so worn-out you could rotate the key halfway and turn off the engine, then remove it. To start it up again, you didn't need a key at all. Just grab the ignition and twist. It was transportation well suited for a bunch of impatient guys who half the time couldn't find the ignition key anyway.

But despite the horseplay and silliness, it was becoming apparent that Thefacebook was turning into a serious business. Zuckerberg knew he had to take more deliberate steps to keep it evolving both technologically and as a business. That summer the growth started to seem a bit scary. They didn't add any new schools until midsummer, but membership kept steadily climbing all summer at the thirty-four colleges where Thefacebook was already operating. And everybody assumed the beginning of the school year would bring massive new demands. New users meant they needed more reliable software and more computing power.

The software and data for Thefacebook was running on servers at a shared facility in Santa Clara, twelve miles south. The guys had to drive down there frequently to unbox, install, and wire up more servers—an activity for which they often recruited friends to help.

They began assuming that Thefacebook was going to continue to keep growing. Every time the database was upgraded or the server array reconfigured, Zuckerberg tried to do it in a way that could accommodate ten times more users than Thefacebook had at that moment. This implicit optimism proved incredibly prescient. If Zuckerberg hadn't had that confidence as early as the summer of 2004, his company might have easily suffered embarrassing and possibly catastrophic outages. But the specter of Friendster's failure to manage its own growth loomed large. Zuckerberg was determined it would not happen to Thefacebook.

The twenty-year-old CEO became obsessed with how well Thefacebook was working technically. He knew that for a communications service like this, performance was key. If the speed with which it delivered new pages to users began to slow, that could be the kiss of death—the beginning of being "Friendstered." There had already been a few frightening outages and slowdowns. He and Moskovitz inserted a timer in the software that discreetly showed on every page just how long the servers had taken to display it. He would argue with the others if they proposed a feature that might reduce that speed. Milliseconds mattered. In an article published around this time, Zuckerberg was quoted saying, "I need servers just as much as I need food. I could probably go a while without eating, but if we don't have enough servers then the site is screwed."

But there was an additional factor that helped spare Thefacebook from performance disaster in its early days, even as its users' zeal and number continued to shock its founders. Zuckerberg and Moskovitz were able to deliberately pace Thefacebook's growth. They did it by deciding when to turn on new schools. Traffic growth followed a clear pattern— launch at a new school and watch usage build steadily, then level off. Each time they added a campus, traffic surged. So if the systems were acting up, capacity was at the max, or they couldn't yet afford new servers, they'd simply wait before launching at the next school. This was a rare asset in an underfinanced Web start-up. It allowed Thefacebook to grow methodically even though it was being run by a bunch of inexperienced kids. Says Zuckerberg: "We didn't just go out and get a lot of investment and scale it. We kind of intentionally slowed it down in the beginning. We literally rolled it out school by school."

Another key factor in Thefacebook's early success was its use of open-source software. From the beginning its database was the open-source MySQL. It cost nothing, nor did PHP, the special programming language for website development that governed how Thefacebook's pages worked. In fact, an up-from-the-bottom Web business like this without real backers could not have emerged much before this. Open-source Web operations software in 2004 had only recently achieved robustness and maturity. Without it, Zuckerberg would not have been able to create a fully featured website in his dorm room and pay for nothing other than the server to run it. Even with 100,000 users, the company's only real costs were the servers and salaries.

Nonetheless, keeping it all running and buying new equipment as Thefacebook grew was starting to cost real money. Zuckerberg spent about $20,000 in the first couple of weeks his crew was in Palo Alto, mostly to add servers at the hosting facility. And more spending was clearly going to be necessary.

The money came out of the account Saverin had set up in Florida. In addition to the cash he and Zuckerberg had deposited, the account was augmented with a considerable amount of advertising income. But with school out, ad sales had pretty much stopped for the summer.

Parker and the new lawyer were trying to straighten out the company's legal status. The limited-liability corporation Saverin had set up was not a sufficient formal structure. It lacked governing documents to define how the company operated. There were no contracts, no official employees, and no payroll. Outside investment would soon be needed, but to get it Thefacebook would have to be turned into a real company.

However, Saverin started to make that very difficult. By mid-July, Parker was starting to talk to investors about putting money into Thefacebook. But when Saverin got wind of these discussions, he wrote a letter to Zuckerberg saying that the original agreement between the partners was that he would have "control over the business," and he wanted a contract to guarantee him that control. Says Parker: "It was so sophomoric. He fundamentally didn't appreciate the importance of product design and technology in this picture. He had this idea that the business stuff was what was important and all this product design and user interface design and engineering and code—you just hire a bunch of engineers and put them in the engine room and they take care of that, you know?" The product as it is engineered and programmed and designed is the business for an Internet company, especially a nascent one. The slightest strategic error in advancing and operating those could mean there would be no more ads to sell.

Whether or not Saverin understood the essential mechanics of launching an Internet company, there were good reasons for him to feel frustrated with the Palo Alto crowd. He had invested his own money (or his family's) and he was the guy working with Y2M and making the calls to bring in ads. Meanwhile, he felt his partner was blasé, to say the least, about revenue. When there was a request for some special treatment from an advertiser, Saverin would bring it to Zuckerberg and Moskovitz. He frequently met a brick wall. What was the chance his investment was ever going to amount to much if Thefacebook couldn't be turned into a proper business? Zuckerberg seemed content that there merely be enough money to pay the bills and keep the site operating.

Saverin had a difficult job at Thefacebook. Advertisers demand responsiveness. They want recipients of their money to be available if they have a question or problem—usually immediately. It was thus harder for Saverin to set his own hours as Zuckerberg and Moskovitz could. His job, unlike theirs, required interacting with customers. It wasn't easy to do that and still keep up with his courses at Harvard.

But he did share one thing with Zuckerberg—ambivalence about Thefacebook's likelihood of future success. He made no secret that Thefacebook was just one of his business activities. He planned to enroll in business school after graduation, so keeping his grades up mattered, despite whatever the company might want from him.

All this later led to a lawsuit. In a legal filing, Zuckerberg and company characterized Saverin's position: "Until he had written authority to do what he wanted with the business, he would obstruct the efforts of the other shareholders and the advancement of the business itself. Saverin also stated that since he owned 30% of the business, he would make it impossible for the business to raise any financing until this matter was resolved."

As their disagreements sharpened, Zuckerberg and Saverin had endless phone calls, which seldom ended with any clear resolution. The Palo Alto group took the view that Saverin was pushing so hard mostly because his father, the hard-driving, self-made Brazilian multimillionaire, was urging him to. "His father was telling him to play hardball," says Parker, "but this is not somebody who should be playing hardball." Parker reports that when pressed to make a decision about something, Saverin would often say either "I have to go talk to my dad" or "I can't give you an answer now." A day or two later he would predictably come back with a firm answer—one that was unyielding.

Despite his hardball, everybody still liked Saverin. He was charming and congenial and smart. But since he didn't seem to be making a commitment to the company like the rest of them, his efforts to get more authority didn't make sense. He was, in effect, demanding to be CEO of Thefacebook without even making a full-time commitment. The boys were inexperienced, but they were working hard, usually until all hours every night, doing whatever had to be done. Saverin appeared to be luxuriating in New York. He didn't get it, they thought.

In any case, Saverin's business skills didn't impress his colleagues. Saverin was getting a lot of business from Internet banner ad networks that bought space in bulk, but they paid very little, and would take months before they did pay. Even Tricia Black, who has a higher opinion of Saverin than the co-founders had, acknowledges that "there were situations where there wasn't any follow-through or there were problems with advertisers."

When Saverin had an idea for Thefacebook, it didn't always go over well with his colleagues. For instance, he thought it would be smart to change the process for requesting a new friend so that it required an additional mouse click. To Zuckerberg—fanatically devoted to making his service easy to use—that was apostasy. But Saverin thought it made sense because in the interim you could show the user an additional ad. There couldn't be a worse reason to do it, in Zuckerberg's opinion. Saverin argued strenuously with Zuckerberg and Moskovitz that Thefacebook ought to put a big banner ad at the top of the page. "We just thought that was the worst possible thing you could do," says Moskovitz. "We thought we would make more revenue in the long run if we didn't compromise the site."

Parker and the lawyer, meanwhile, were preparing to create an entirely new legal structure. They were filing papers to incorporate Thefacebook in Delaware. (Most American companies—including just about all Silicon Valley start-ups—incorporate there because Delaware's laws are favorable for business.) Parker, managing the restructuring, was particularly concerned that the intellectual property (IP) that defined what Thefacebook was—that is, the company's most critical possession—was not owned by the company. Saverin in setting up the LLC had not sufficiently defined what it controlled. (As the creator, most of the software and design was by rights owned by Zuckerberg personally, along with some owned by Moskovitz.) Legally speaking, there hardly was a company before this point. Saverin controlled the bank account, but the servers where the service actually resided, along with the intellectual property, were under the control of Zuckerberg, Moskovitz, and Parker. The Florida LLC was more or less an empty shell, and what it actually owned was unclear. Zuckerberg and Moskovitz signed over their portion of the LLC, plus the critical IP, to the new Delaware corporation.

Zuckerberg won't talk about this dispute now, but his legal filings say he told Saverin that because he refused to move to California with the rest of them and had not done work he'd said he'd do, he would subsequently no longer be an employee of the company. While his ownership interests would remain, they were inevitably subject to dilution (meaning they would represent a smaller and smaller percent of the total company) as employees were hired and given stock options, and as investors bought into Thefacebook. Zuckerberg and Moskovitz, by contrast, would be eligible to receive additional grants of stock based on their continuing contributions.

The new corporate bylaws provided that Zuckerberg, with 51 percent ownership, was the company's sole director. Saverin got 34.4 percent. Zuckerberg upped Moskovitz's portion of the company to 6.81

percent in recognition of his increasing contributions. He also gave
his new confidant Parker 6.47 percent. But apparently nobody's loyalty
could be taken for granted at this point, so the shares of both Parker
and Moskovitz were to double if they stayed until the following year,
which would significantly dilute Saverin's share. Interviewed by the
Harvard Crimson a few months later, Zuckerberg explained why he'd
increased Moskovitz's stake: "Everyone else was like, 'What the fuck
are you doing?' And I was like, 'What do you mean? This is the right
thing to be doing. He clearly does a lot of work.'" The law firm got the
remaining 1.29 percent.

Saverin later claimed that he did not know the company was being
reincorporated, or about several other aspects of this plan. But something he learned around this time must have made him a lot angrier
because this is when he "attempted to hijack the business," in the
words of Thefacebook's later legal filing. He froze the Florida bank account, making it impossible for the company to pay its bills. He said he
wouldn't release any money until his business demands were met. "It
felt like we were negotiating with terrorists," says someone who was in
the Palo Alto house. This was just when it had become apparent that big
purchases of new servers would soon be required. Saverin said he had
prepared an operating agreement that described the respective roles the
boys would play in the company, but he wouldn't let Zuckerberg see
it unless he promised to sign it without showing it to his lawyer or anybody else. Zuckerberg responded by creating his own document, which
described the responsibilities he believed were appropriate for both of
them, but Saverin would have none of it.

As the negotiations continued, Zuckerberg had to spend his own
money to keep the lights on at 819 La Jennifer Way and more importantly, to keep buying servers. Zuckerberg had tens of thousands of dollars he had saved from programming and website jobs he'd done in his
summers and spare time. His dentist father and psychologist mother
also contributed thousands. This was money, according to a later lawsuit, that had been intended for his college tuition. Zuckerberg and his
family ended up spending $85,000 that summer. For twenty-five new
servers alone, he spent $28,000.

Chris Hughes didn't return from France and show up at the house until the end of the summer. But even so, he played a critical part in Zuckerberg's brain trust. Thefacebook's Palo Alto geeks lacked confidence in their own judgments about how people would respond to the product. Humanities major Hughes had a better feel than they did for how users would respond to new features. Immediately upon his arrival Hughes was deluged with requests to look at this or that feature or page design. He talked a lot about privacy and simplicity. Even after Hughes left to go back to school for his junior year, master and commander Zuckerberg often wielded Hughes's opinions when arguing a point with one of the others. Hughes remained Thefacebook's public spokesman, fielding an ever-growing number of interview requests out of his dorm room, mostly from college papers around the country.

At summer's end, Thefacebook had over 200,000 users. Zuckerberg and Moskovitz were planning to launch at seventy more campuses in September. Parker was well along in continued discussions with investors who the guys hoped would give them the money they needed without too many strings. Negotiations with Saverin continued.

Some weeks earlier, Zuckerberg and Moskovitz took about five minutes to decide they wouldn't return to Harvard. Earlier they had thought they'd be able to run Thefacebook from their dorm room again, but signs were that this could be an explosive school year for the service. They didn't want to mess it up. D'Angelo and the interns returned to school, as did Saverin. Zuckerberg, Moskovitz, Parker, and Halicioglu were, for now, Thefacebook. McCollum stayed on to work on Wirehog.

On September 11, the owners of the house stopped by to check on its condition. They did not like what they found. Zuckerberg had sublet it from tenants for the summer. In a later court case, a memo the owners subsequently wrote was entered in the record. "The house appeared to be in total disarray and very dirty," they wrote. "Furniture out in garage—unsure about what is missing and/or broken . . . Ashes from bar-b-q dumped—some on deck and some in a flowerpot out in back yard. Broken glass all around yard and some on deck . . . An antique Indian basket . . . had been taken outside and left on top of the built-in bar-b-q. It was broken and burned. . . . " They also complained about damage to the chimney from the zip line, repair costs for the pool filter damaged by pieces of glass, a broken laundry room door, etc. College shenanigans had been extensive at Thefacebook's corporate headquarters.

In early September, even as he was still wrestling by phone with Saverin, Zuckerberg was served documents informing him that Tyler and Cameron Winkelvoss and Divya Narendra had filed suit in federal court. They contended that Zuckerberg had stolen the idea for Thefacebook from them.

3

Social Networking and the Internet
"Every capitalist out there wants a piece of the action."

The concepts of social networking are not new, and many of the components of the early Facebook were originally pioneered by others. Zuckerberg has been accused several times of stealing ideas to create Facebook. But in fact his service is heir to ideas that have been evolving for forty years.

Something like Facebook was envisioned by engineers who laid the groundwork for the Internet. In a 1968 essay by J. C. R. Licklider and Robert W. Taylor titled "The Computer as Communication Device," the authors asked, "What will on-line interactive communities be like? In most fields they will consist of geographically separated members, sometimes grouped in small clusters and sometimes working individually. They will be communities not of common location, but of common interest." The article crept further toward the concept of social networking when it said, "You will not send a letter or a telegram; you will simply identify the people whose files should be linked to yours." As a key employee in the Advance Research Projects Agency of the Department of Defense, Licklider helped conceive and fund what became the ARPAnet, which in turn led to the Internet.

A decade or so later, a few pioneers were beginning to spend time in such online communities. The first service on the Internet that captured substantial numbers of nontechnical users—long before the invention of the World Wide Web—was the Usenet. Begun in 1979, it enabled people to post messages to groups dedicated to specific topics. It functions to this day. In 1985, Stewart Brand, Larry Brilliant, and a couple of others launched an electronic bulletin board called The Whole Earth 'Lectronic Link, or Well, in San Francisco. In 1987, Howard Rheingold, a big user of the Well, published an essay in which he coined the term *virtual community* to describe this new experience. "A virtual community is a group of people who may or may not meet one another face to face," Rheingold wrote, "and who exchange words and ideas through the mediation of computer bulletin boards and networks."

More and more people became familiar with electronic communication, initially by commenting in online groups and chat rooms. The French postal service was the first to bring these concepts to a mass consumer audience when it launched a national online service there called Minitel in 1982. Then America Online started in 1985, initially under another name. In 1988, IBM and Sears created an ambitious commercial online service called Prodigy. Shortly, however, AOL came to dominate the business in the United States. On these services people typically invented or had assigned a quasi-anonymous username for themselves, which they used for interacting with others. I was Davidk4068 on AOL. By the early 1990s, ordinary people began using electronic mail, again typically using addresses that did not correspond to their names. Though they maintained email address books inside these services, members did not otherwise identify their real-life friends or establish regular communication pathways with them. Later in the decade, instant-messaging services took hold the same way—people used pseudonymous labels for themselves, not their names.

In the early days of the World Wide Web, the notion of an online community advanced a little further. Services like TheGlobe.com, Geocities, and Tripod emerged and enabled users to set up a personal home page that could in some cases link to pages created by other members. Mark Zuckerberg's first website was one he created on Geocities when still in junior high school. The popular fee-based dating site Match.com launched in 1994 filled with personal information, but for a very specific purpose. Classmates.com debuted in 1995 as a way to help people, identified by their real names, find and communicate with former school friends.

The era of modern social networking finally began in early 1997. That's when a New York–based start-up called sixdegrees.com inaugurated a breakthrough service based on real names. Two Internet sociologists, danah boyd and Nicole Ellison, articulated in a 2007 paper the salient features of a true social network: a service where users can "construct a public or semi-public profile," "articulate a list of other users with whom they share a connection," and "view and traverse their list of connections and those made by others within the system." You establish your position in a complex network of relationships, and your profile positions you in the context of these relationships, usually in order to uncover otherwise hidden points of common interest or connection. Another element must be added to explain the trends that led to Facebook—an online profile based upon a user's genuine identity.

The sixdegrees service was the first online business that attempted to identify and map a set of real relationships between real people using their real names, and it was visionary for its time. Its name evokes the speculative concept that everyone on earth can be connected through an extended chain of relationships that begins with your immediate friends, proceeds to the next "degree"—the friends of your friends, and on until the sixth "degree."

Andrew Weinreich, sixdegrees' founder and a lawyer, was himself an inveterate networker. The World Wide Web was just beginning to get traction among ordinary people. At sixdegrees' launch in early 1997, Weinreich invited several hundred people assembled at New York's Puck Building to join immediately at one of the twenty PCs set up there in the room. "It no longer makes sense for your Rolodex to live on your computer," he proclaimed. "We'll place your Rolodex in a central location. If everyone uploads their Rolodex, you should be able to traverse the world!"

Members normally joined sixdegrees after receiving an email invitation from an existing member. This method of recruitment would be imitated by many subsequent social networks. It sounds obvious to us now, but at the time it was revolutionary. The service allowed you to create a personal profile listing information about you and your interests, based on your real name. Then it helped you establish an electronic connection with friends. You could search through profiles and ask friends to introduce you to interesting people you found. There were two key features on sixdegrees when it launched. The first was "connect me." If you put in someone's name, it would create a map of your relationship to them through the service's various members. The other was "network me," which enabled you to identify certain characteristics you were looking for, so the service could identify members who matched those qualities. A doctor in Scarsdale who likes chess, perhaps?

But, as Weinreich now ruefully concedes, "We were early. Timing is everything." The service was hugely expensive to develop and operate. It hired ninety employees, bought lots of expensive servers and database software licenses from Oracle, and paid Web-development firms Sapient and Scient millions to develop features. And what did all this expense make possible? A service that most people used at a painfully slow speed on a dial-up modem. And there were other severe limitations. Profiles may have had your name, career data, and favorite movies, but they lacked photographs. After all, few people back then had digital cameras. The lack of photos was such an obvious problem that in 1999 Weinreich seriously considered asking members to send in photo prints of themselves so interns could upload them assembly-line style.

It was unclear to people—members and nonmembers alike—whether sixdegrees was intended as a dating service, a business networking service, or both. Nonetheless, by 1999 sixdegrees had reached 3.5 million registered users and a larger company bought it for $125 million. But it never generated much revenue, and in the wake of the dot-com bust its new owner shut the money-losing company down in late 2000. Figuring that this was the beginning, not the end, of social networking, lawyer Weinreich and his partners had the foresight to win a very broad patent covering sixdegrees' innovations, a patent that would later figure in the history of Facebook. Weinreich talked about networks like his as the "operating system of the future."

Though sixdegrees broke the ice, it took years before others ventured into the waters and built what can be considered genuine social networks. In 1999, two ethnic-focused sites, BlackPlanet and Asian Avenue, launched with limited social networking functions. A Swedish social network for teenagers called LunarStorm launched January 1, 2000. Cyworld, a hugely popular service in Korea, added social networking capabilities in 2001.

. . .

It wasn't until 2001 and 2002 that the social networking bug hit Silicon Valley and San Francisco. Most of the entrepreneurs and venture capitalists there were still in shock following the precipitous slide in valuations and revenues for Internet companies that began in early 2000. Companies were closing and the mood was grim, especially for consumer Internet companies. New ones were hardly receiving any investment money in 2001 and 2002. But a few hardy souls recognized that sixdegrees might have simply started too soon.

Plaxo, the Internet company that Sean Parker founded with friends in 2001, wasn't a social network, but it had a lot in common with them. Plaxo was a contact management service. After new members uploaded their contacts, it relentlessly peppered those people with requests to update their information, always pressing them to join as well. It was obnoxious, but it worked often enough. Parker was thinking much like Andrew Weinreich at sixdegrees—put your Rolodex in a central location and let us manage it for you. Parker liked the Plaxo concept because it was viral—one user could lead to an entire chain of users. Plaxo also foreshadowed a crucial aspect of Facebook—it maintained unique identifying information for individuals based upon that person's network of contacts.

In late 2001, an entrepreneur and local pioneer named Adrian Scott launched a social network called Ryze. Scott aimed to eliminate any uncertainty about Ryze's purpose. It was not a dating site. It was for businesspeople. Its name was intended to evoke the way members could "rise up" by improving the quality of their personal business network. Members' profiles focused on work accomplishments and they networked with co-workers and business contacts. It planned to make money by charging employers and others to search its databases for prospective employees, consultants, etc. Though it never much took hold except among San Francisco's tech cognoscenti, it inspired and set the stage for many developments that followed.

Jonathan Abrams, a local programmer, Ryze member, and inveterate partyer, saw an opportunity to focus on the nonwork part of people's lives. He built a very social network for consumers and called it

Friendster. Though it wasn't exactly a dating site, it offered many tools to help members find dates. Abrams gambled that he could take customers away from Match.com, as the idea was that you'd meet more interesting people if you got to know the friends of your friends. Members were expected to use their real names, and Friendster gave you a novel tool to keep track of people—the very one that sixdegrees' Weinreich had pined for. Their pictures appeared next to their names right on their profile. This was a breakthrough. You could search to learn which people lived near you who were already friends of a friend. If you liked their picture, you could try to connect.

When Friendster launched in February 2003 it was an immediate hit. Within months it had several million users. To join you needed an invitation from an existing user, which were much in demand. Pretty soon people were talking about Friendster as the "next Google." It even reportedly turned down a $30 million buyout offer from Google itself. Back in Boston, Mark Zuckerberg took notice and joined, as did other Harvard undergraduates, including the Winkelvoss twins.

Friendster seemed to be hitting the big time. Abrams made magazine covers. But by the middle of the year the experience of users started spiraling downward. Millions were joining and Friendster's servers were slowing. It couldn't manage its success. Pages took twenty seconds to load. It also started to have public relations troubles—it engaged in a very public battle with so-called "fakesters," users who were deliberately creating Friendster profiles using phony names and identities, including cartoon characters and dogs. Abrams was resolute that people on Friendster should use their real names, and he kicked lots of the fakesters off. Aiming in part to solve its expensive technical challenges, the company took a big infusion of money in fall 2003 from two eminent venture capital funds—$13 million from Benchmark Capital and Kleiner Perkins Caufield & Byers.

A recent visit to the San Francisco office of Friendster founder Abrams, now operating an online invitations business called Socializr, finds him scruffy-bearded, contrite, and still eager to party. The first thing he does when I walk in is offer me a tequila. He interrupts the interview several times to reiterate the offer, despite my repeated refusals. "The site didn't work well for two years—that's a fact," he concedes, finally getting down to business. Then he explains how a series of engineering misjudgments prevented Friendster from repairing its performance problems until long after he was removed by the investors as CEO in March 2004.

Abrams is one of social networking's great innovators, but he willingly concedes he built on the ideas of others. "The concepts were not new," he says. "What was new was the vibe of it, the design, the features." But the fact is, as Sean Parker puts it, "Jonathan cracked the code. He defined the basic structure of what we now call a social network."

In Friendster's wake, a throng of social networking sites blossomed in San Francisco attempting to duplicate its appeal. Each tackled the idea of connecting people in a slightly different way. One was Tickle, a service which, on observing Friendster's broad-based appeal, altered its own service, which had previously been based on self-administered quizzes and tests. Two of the other new social sites—LinkedIn and Tribe.net—were founded by friends of Abrams.

Reid Hoffman had been the lead angel investor in Friendster's very first financing, putting in $20,000 of the total $100,000 Abrams raised. Hoffman is a pivotal figure in the history of social networking. One of Silicon Valley's most thoughtful executives, he carries a substantial amount of industry credibility in his stout frame. Way back in August 1997 he started a dating service called SocialNet, which tried to find matches based on information users put in a profile. Some call it the very first social network, though Hoffman doesn't. In any case, it didn't do very well as a business (though when it was sold, its investors made their money back). But in May 2003, three months after the launch of Friendster, Hoffman founded LinkedIn, a social network for businesspeople. Hoffman believed that social networking was likely to divide into two categories—personal and business—so this was not a conflict with his support of Friendster. LinkedIn, which thrives to this day, has a lot in common with Ryze. Your profile is basically your résumé. Users look for jobs and ask others for business recommendations or advice. But in keeping with its businesslike attitude, it started without photos. (Hoffman added that capability later.)

Mark Pincus plays Laurel to Hoffman's Hardy. Skinny, medium-height, and hyperactive, Pincus was another Friendster investor and Hoffman's buddy. In May 2003, the same time that Hoffman was launching LinkedIn, Pincus unveiled Tribe.net, a social network where members could create a "tribe" around a specific interest. Tribe.net was originally intended to help members share Craigslist-type classifieds so they could buy things from people they knew. Its tribal quality, however, quickly became its trademark, and its most cohesive online tribes were not the ordinary-Joe ones Pincus had envisioned. They included regular attendees at the annual Burning Man festival in Nevada as well as devotees of alternative sexual practices, more interested in just connecting than in buying and selling things.

Sean Parker fell in with this San Francisco social networking mafia. At that time, Parker shared a house with Stanford students in Palo Alto, where Friendster was already taking off, and several of these guys were already in his own real-world social network. Ryze's Adrian Scott had been an early Napster investor. And Tribe's Pincus had founded Free-loader, the Arlington, Virginia, start-up where Parker interned in 1994 at age fifteen. Soon Parker was hanging out with them and their friend Abrams of Friendster.

Parker and Abrams quickly bonded. And the more he hung around Abrams, the more Parker became fascinated by Friendster. He began spending lots of time at the Friendster offices. He helped Abrams find additional investors and acquired a small amount of Friendster stock himself. That was just when the service started to bend and break under the load of its newfound popularity. "I watched from the sidelines as they lost the war," says Parker now. "The story always was: 'One more month, one more month. We'll get it working.'" (Friendster was later resuscitated, but too late for the U.S. market. Now about 60 percent of users are in the Philippines, Indonesia, and Malaysia.)

In the summer of 2003, just as Tribe.net and LinkedIn began to grow, an unexpected development got Pincus and Hoffman worried. They learned that the patent held by the now-extinct sixdegrees was being put up for auction by its new owners. The patent is broad and sweeping. It describes a social network service that maintains a database, enables a member to create an account, then encourages him or her to invite others to connect to their network via email. If this other person accepts the invitation and confirms his friendship, the service creates a two-way communications connection. These processes are at the heart of most social networks.

Lawyers for the two entrepreneurs told them that in the wrong hands the patent could be used to stop both of their companies, or pretty much any social networking firm. They decided to try to buy it. They also knew that Friendster was getting millions of dollars from the VCs, and worried that with more resources Friendster might attempt to elbow into their parts of the industry. Owning the patent was a form of defense. However, neither company's board would authorize buying the patent. So Hoffman and Pincus decided to use their own money.

But they weren't the only ones who had recognized the patent's potency. Yahoo was beginning to realize it might have missed the social networking boat. It entered the auction and actually put in the highest bid. But Hoffman and Pincus, gung-ho, were willing to pay faster and won with a bid of $700,000.

The two now say they just wanted to keep the patent out of the hands of larger players like Yahoo or Friendster. "We were worried that someone was going to buy the patent and then sue all the early social networking companies," says Hoffman now. "We bought it defensively, to make sure no one would kill the nascent industry."

But even as these entrepreneurs were creating a new industry in San Francisco, an unlikely competitor emerged from nowhere, four hundred miles south in Los Angeles. MySpace began as one of scores of Friendster clones—MySpace co-founder Tom Anderson was an avid Friendster user. Anderson got the idea to start MySpace in part out of frustration as Friendster slowed and crashed. But, according to *Stealing MySpace*, the definitive history of MySpace, by Julia Angwin, Anderson also thought he could deliberately appeal to the so-called fakesters, "to create a site where users could create any identity they liked." He and co-founder Chris DeWolfe put few restrictions on how you could use MySpace.

The two were employees of an unwieldy and disorganized Net conglomerate called eUniverse, which secretly installed spyware on users' PCs and sold expensive and questionably advertised merchandise. There they applied their libertine values to the creation of the new service. Anderson and DeWolfe took a kitchen-sink approach. If something had proven popular on the Web, the commercially minded pair wanted it in MySpace. When their service launched on August 15, 2003, only six months after Friendster and three months after Tribe.net, it included games, a horoscope, and blogging along with a Friendster-like profile page for members.

While Friendster's Abrams was a bit of a control freak, fighting a lengthy losing battle to protect his particular vision of a real identity-based service, MySpace took a generally lax approach to just about everything. That suited members just fine. For one thing, it was less rigid about who could join than other social networks. You didn't need an invitation from an existing member. You could use either a real name or a pseudonym. And one of the features members liked best was not even intentional. An initial programming error allowed members to download Web software code—called HTML—onto their profiles. People quickly began using it to tart up their sites. Ever adaptive, the MySpace founders noted members' enthusiasm for this freedom and embraced the error as an asset.

Member-created designs were how MySpace got its distinctive Times Square look—all flashing graphics and ribald images. But while this look might have been unintentional, it was in keeping with the MySpace ethos—if you could pretend to be anybody, you also had the freedom to make your profile look like anything. And you didn't even always know who a MySpace member was. That made it difficult to limit your connections to genuine friends. People began adding friends willy-nilly, the more the better. It became a competition—how many could you have? As for behavior on the site, along with plenty of conventional conversation there was a definite tilt toward the sexual. On Friendster the look of a profile was fixed for consistency and Abrams wanted you to use your real name for connecting to other real people. Such niceties were disregarded by MySpace's Anderson and DeWolfe.

As Angwin carefully explains in *Stealing MySpace,* the canny founders had superb timing. The world was ready for a mass-market social network. The sixdegrees service came too soon—it lacked the right online environment in which to thrive. But that landscape had finally emerged. In 2003, Angwin notes, the percentage of Americans with broadband Internet access rose from 15 percent to 25 percent. Broadband not only meant faster viewing times, it also made uploading photos easier. Digital cameras were becoming common and affordable. Crucially, a wider variety of people were getting fast Net speeds. For the first time lots of families—including those with teenage girls—had broadband. Had Friendster not broken down under the strain of success, it might have appealed to this crowd, but MySpace nicely stepped into the void.

MySpace initially spread among the relatively hip Los Angeles friends of Anderson and DeWolfe. The founders marketed their service in clubs to both bands and audiences. Shortly afterward it became an essential promotional tool for bands in L.A. It didn't take long before enterprising musicians all over the country began adopting MySpace. Along with the bands came the bands' audience—teenagers.

MySpace was hip and a great site to find out about bands, but it also leaned toward the sexual. Holding MySpace parties in nightclubs around the country became another of the site's promotional tools. The implicit message: MySpace was a digital club where wild behavior was welcome. A disenchanted Friendster user who called herself Tila Tequila joined MySpace, bringing her fan club with her. She was a buxom young Vietnamese model with a yen for attention. Her profile was full of pictures of her wearing very skimpy clothing.

Though the site's minimum age was supposedly sixteen, plenty of younger kids created profiles claiming to be older. It wasn't unusual for thirteen-year-old eighth-grade girls to post photos of themselves wearing only a bra. Parents groups at junior highs and high schools all over the country convened alarmed meetings about the dangers of social networking.

By the time Thefacebook launched in February 2004, the flamboyant MySpace had more than a million members and was quickly becoming the nation's dominant social network. Thefacebook offered users limited functions, a stark-white profile page, and was limited to students at elite universities. The contrast in tone couldn't have been greater.

The first social network explicitly intended for college students had begun at Stanford University in November 2001. It was probably also the first real social network ever launched in the United States. This little-known service, called Club Nexus, was designed by a Turkish doctoral student in computer science named Orkut Buyukkokten as a way for Stanford students to improve their social life. An undergraduate political science major named Tyler Ziemann managed the nontechnical parts of the project.

Club Nexus was revolutionary and had a raft of features — probably too many. It allowed members to create a profile using their real names, and then list their best on-campus friends, who were known as "buddies" in Club Nexus lingo. Buddies who were not already members then automatically received an email inviting them to join the service. Only students with a Stanford-issued email address could join, and that email authentification ensured that each person was who they said they were. You could chat, invite friends to events, post items in a classifieds section including personal ads, write bloglike columns, and use a sophisticated search function to find people with similar interests. Students used it to find study partners, running buddies, and dates. Buyukkokten himself once bragged that what made it different from any other website was "you can create really big parties."

Within six weeks Club Nexus had 1,500 members at Stanford, whose student body totaled about 15,000. But after it reached about 2,500, usage leveled off. The service was just too complicated. Buyukkokten was a talented programmer who had loaded it with every interesting feature he could think of. But that made it difficult to use and diffused activity among many different features. You didn't get the sense there were many others in there with you.

Once the two men got their degrees in 2002, they wanted to commercialize their venture. Recognizing that student use was tepid, they made what some might call a foolish decision, considering the subsequent successes of Facebook—they focused instead on alumni. They created a company called Affinity Engines, which began marketing a modified version of Club Nexus called InCircle to college alumni groups. Their first client was the Stanford Alumni Association. By 2005 its customers included alumni networks at thirty-five schools, including giants like the University of Michigan. But not long after Affinity Engines started, Orkut Buyukkokten left the company and went to Google.

A year or so after he joined Google, the entrepreneurial programmer approached Marissa Mayer, a top company product executive, and told her that over the weekend he'd built the prototype for a new social network. Mayer and Google's executives, who by policy encourage entrepreneurship among employees, embraced his project. Google was thinking of calling the project "Eden" or "Paradise." Then one day Adam Smith, a product manager working with Buyukkokten, told Mayer that the engineer owned the Web address Orkut.com. The two felt Buyukkokten embodied the spirit of his service, so they just decided to name it after him.

The well-conceived Orkut, a social network open to anyone, launched in January 2004, just two weeks before Thefacebook.com. It initially thrived in the United States and was holding its own against a surging MySpace. But by the end of 2004 it had, somewhat oddly, been tightly embraced by Brazilians. A grassroots campaign there to win more members than Orkut had in the United States captured the imagination of young Brazilians. After they succeeded, the service acquired a distinctly Brazilian and Portuguese-speaking cast. Americans began to drop away. Today Orkut, still owned by Google, remains one of the world's largest and most sophisticated social networks, yet more than half its membership is Brazilian. Another 20 percent live in India. Google's diminished expectations for it can perhaps be gleaned from the fact that in 2008 it moved Orkut's headquarters to Brazil.

Club Nexus was the first college-specific social network, but by the 2003–2004 school year similar sites were popping up at a number of schools. The Daily Jolt, a sort of discussion community, had been around since 1999 as a kind of campus bulletin board and was operating at twelve schools. Collegester.com—"a virtual community of free, useful, and enjoyable services 'for the students, by the students'"—had been launched in August 2003 by two University of California at Irvine alumni. An online matchmaking service called WesMatch was thriving at Wesleyan University. Its entrepreneurial founders had launched a version at Williams College and were expanding to Bowdoin, Colby, and Oberlin. At Yale, the student-run College Council launched a dating website called YaleStation on February 12—only a week after Thefacebook's debut. By the end of the month about two-thirds of undergraduates had registered. Then there was CUCommunity, which had taken off at Columbia in January. Both the Yale and Columbia sites were getting lots of members before Thefacebook arrived.

In late 2003 the Ivy League seems to have collectively decided that campus facebooks should go online. Student governments at Cornell, Dartmouth, Princeton, Penn, Yale, and Harvard, among others, were all complaining to college administrations that their campus facebook was not in digital form. The idea was no secret. A sense that the time had come helped push Zuckerberg to create Thefacebook and accounts for its name. Students everywhere had also been influenced by the rapid ascendancy of Friendster, and many were dismayed to see it stumbling. By fall, MySpace was already making waves in Los Angeles and in the music world.

Aaron Greenspan, a Harvard senior, launched a service there in September 2003 called houseSYSTEM. It allowed residents of Harvard residential houses to buy and sell books and to review courses, among other functions. It also invited students to upload their photographs to something called the Universal Face Book. houseSYSTEM was controversial for how it treated student passwords and never got much usage, though hundreds of students signed up to try it.

Separately, Divya Narendra claims to have come up with the idea for a Harvard-specific social network in December 2002. He later teamed up with the Winkelvoss brothers to build Harvard Connection, according to some of the voluminous legal documents filed in the lawsuit they brought against Zuckerberg and Facebook. The identical towering Winkelvoss twins—known to some Harvard classmates as "the Winklevii"—worked hard at rowing for years and made the finals for men's pair rowing at the 2008 Olympics in Beijing. They came in sixth and last, but it was a huge achievement. Previously they'd won a gold medal at the Pan-American Games in Rio de Janeiro. The two athletic blond uber-WASPs couldn't be more different from the scrawny, nerdy, brainy Jews who founded Thefacebook.

The three worked fitfully on the idea that would become Harvard Connection over the next year. Since none of them were programmers themselves, they hired people to help. Two successive computer science students tried and failed, in the founders' opinion, to get Harvard Connection right.

During the first months of Harvard's fall 2003 semester, Zuckerberg began making waves with ad hoc bad-boy applications that were intrinsically social—first Course Match, then Facemash. Narendra and the Winkelvosses read about him in the *Crimson*'s coverage of the Facemash episode. They got in touch and arranged a meeting. He agreed to help out, but says now he thought of it as just another of his many social software "projects."

Zuckerberg worked off and on writing code for Harvard Connection. After a few weeks he appears to have lost interest, though he apparently didn't make that clear to the Winkelvosses and Narendra. They began to complain that he was taking too long. At one point Zuckerberg apologized for a delay, explaining he had forgotten to bring the charger for his laptop home with him for the Thanksgiving holidays. Eventually Harvard Connection's trio accused Zuckerberg in a federal lawsuit of stealing their intellectual property. The case was settled in mid-2008 with the requirement that parties not disclose details. But some trial documents became public, including emails between the complainants and Zuckerberg. These exchanges give a picture of what

Harvard Connection was intended to be. In one, Cameron Winkelvoss included proposed text for a page: "Harvard Connection has compiled a list of the premiere nights of the hottest clubs and lounges in the Boston area. We have brokered deals with promoters at these clubs to give all of our registered users reduced admission on these given nights." This sort of discounted partying seems to have been a major focus of the planned site.

On December 6, Cameron Winkelvoss emailed Zuckerberg again: "One idea I came up with is an 'incest rating.' . . . Essentially it is a measure of how close your interests and the person's interests you are looking at are. . . . It would be funny to see how closely related and how 'incestuous' it would be to request a date with a given person." He also suggested the site provide recommendations on who a member should date, and mused that maybe Harvard Connection should deceive users by pretending that these matchups were determined by software algorithm: "Perhaps there could be some random element incorporated into it (obviously people viewing the site shouldn't know this, for all they know it's a thoughtfully calculated recommendation)". Winkelvoss considered himself to be creating, as he put it here, a "dating site."

The emails appear to show that Zuckerberg began avoiding the three Harvard Connection founders. By January 8, 2004, he emailed Cameron: "I'm still a little skeptical that we have enough functionality in the site to really draw the attention and gain the critical mass necessary to get a site like this to run." Yet in late November he had written, "Once I get the graphics we'll be able to launch this thing. . . . It seems like everything is working." The Harvard Connection guys repeatedly requested a meeting. When the four finally convened on January 14, Zuckerberg said he didn't have any more time to work on the project.

Zuckerberg also had a little involvement with houseSYSTEM creator Greenspan. The two met for dinner in early January in the Kirkland dining room. At the meeting, Zuckerberg invited Greenspan to partner with him to create his new project, which he didn't describe in detail. But the older student demurred. In a 333-page self-published, self-justifying autobiography he writes, "I didn't like the idea of working for someone who had just been disciplined for ignoring privacy rights on a massive scale." (He's referring to Facemash.) Greenspan, two classes ahead of Zuckerberg, had run his own small software company since he was fifteen and clearly felt superior to the sophomore.

However, at the same meeting he invited Zuckerberg to incorporate his project, whatever it was, into houseSYSTEM. But Zuckerberg said he didn't want to do that because houseSYSTEM was "too useful," according to the book. Greenspan writes that this statement confused him. "It just does too much stuff," Zuckerberg continued, according to the book. "Like, it's almost overwhelming how useful it is." Today, Zuckerberg won't say much about houseSYSTEM, except that "the trick isn't adding stuff, it's taking away." houseSYSTEM eventually disappeared. Zuckerberg classmate Sam Lessin, himself a programmer and now an Internet entrepreneur, recalls it as "a huge sprawling system that could do all sorts of things." By contrast, he says, Thefacebook was almost obsessively minimal. "The only thing you could do immediately was invite more friends. It was that pureness which drove it."

Thefacebook launched on February 4. Six days after that, Cameron Winkelvoss sent Zuckerberg a letter saying he had misappropriated the Harvard Connection founders' work and owed them damages. The letter demanded Zuckerberg stop working on Thefacebook. Winkelvoss and his partners complained to the Administration Board—the same body that had disciplined Zuckerberg for Facemash. A Harvard dean got involved and asked Zuckerberg for his account of what happened.

In a long letter he wrote the dean on February 17, Zuckerberg said that from his first work on the project, he had been "somewhat disappointed with the quality of the work the previous programmers had done on the site." He called it "messy and bloated." He dismissed the ideas of the Winkelvoss brothers and Narendra. "My most socially inept friends at the school had a better idea of what would attract people to a Website than these guys." He complained about their planning. "I wasn't happy with the way they had failed to come through with their promises of advertising, the necessary hardware to run the site, or even graphics for the site (last time I checked, their front page was still using an image straight out of a Gucci ad)."

"I'm kind of appalled," he continued, "that they're threatening me after the work I've done for them free of charge. . . . I try to shrug it off as a minor annoyance that whenever I do something successful, every capitalist out there wants a piece of the action." He ended his letter about what he called "ridiculous threats" by saying, "I didn't go through the differences between my site and theirs because the two are com-pletely different." The dean decided not to get involved in the dispute.

Were the two services substantially different? As Cameron Winkel-voss's emails indicate, Harvard Connection was intended largely as a party guide and dating service. The intention was to "broker deals with promoters," taking a cut in the process. Thefacebook was noncommercial. It aimed to replace offline facebooks. It was centered on information about individuals. Everything on Thefacebook was generated by its users, while Harvard Connection was going to include content including "reviews of nightclubs."

The Harvard Connection, rebranded ConnectU, finally launched in late spring 2004. That fall, ConnectU's founders, assisted by lawyers who usually worked for the Winkelvosses' affluent father, sued Zuckerberg in federal court in Boston. The lawsuit asserted that Zuckerberg stole ideas including "creating the first niche social network for college/university students"; "serving as a directory of people and their interests and qualifications, a forum for the expression of opinion and ideas, and a safe network of connections"; requiring members to register using their ".edu" email addresses; and launching at Harvard and then extending to other schools, with an eventual plan to include "every accredited academic institution domestically and internationally."

During the time he was working for Harvard Connection, Zuckerberg may have become uneasy about the fact that he was already working on his own social network. He certainly should have alerted the Winkelvoss brothers and Narendra earlier about what to expect. He was rude. He became very uncooperative. But long before he met the Winkelvosses and Divya Narendra he already was thinking about what kind of social software was possible on the Internet. That's why the Harvard Connection project interested him in the first place. The civil lawsuit filed on behalf of the three alleges behavior considerably worse than rudeness: "copyright infringement, breach of actual or implied contract, misappropriation of trade secrets, breach of fiduciary duty, unjust enrichment, unfair business practices, intentional interference with prospective business advantage, breach of duty of good faith and fair dealing, fraud, and breach of confidence." The complainants asked to take over the entire Facebook site and be paid damages equal to its value. Pretty strong stuff for a supposed ten-hour project for which Zuckerberg never had a written contract and was never paid.

Zuckerberg probably refined his own ideas during the course of working on Harvard Connection, but there don't seem to be elements in common between the sites that had not already been used by other services in the past. Every social networking plan on the planet by this time was influenced by Friendster. It did prove critical for Thefacebook to use .edu addresses for registration, but other sites at colleges had already begun taking a similar approach. Club Nexus limited itself to Stanford-only email addresses as early as fall 2001.

In September 2004 when they filed suit against Thefacebook, ConnectU claimed fifteen thousand users at two hundred colleges. Thefacebook competed with it vigorously. ConnectU did lead to one huge success for its founders, however—a 2008 financial settlement of the lawsuit. It gave the ConnectU's creators plenty of money to go away—reportedly $20 million in cash as well as stock in Facebook worth at least $10 million. ConnectU was already moribund, but now it shut down.

Aaron Greenspan also accused Zuckerberg of stealing his ideas. In his autobiography, titled *Authoritas: One Student's Harvard Admissions and the Founding of the Facebook Era*, he writes that he "invented The Facebook while attending Harvard College." In April 2008 he petitioned the U.S. Patent and Trademark Office to cancel the trademark "Facebook." He claimed he was the rightful owner, because he pioneered the name months earlier than Zuckerberg as part of houseSYSTEM. Greenspan served as his own attorney. The Trademark Trial and Appeal Board decided his claims were plausible enough to let the action move forward. Some months later, Facebook settled with Greenspan for an undisclosed sum.

Greenspan doesn't accuse only Zuckerberg. He writes in his book that the Winkelvoss brothers and Narendra took ideas from him, too, and that Harvard Connection was also an imitation of houseSYSTEM.

Social networking has now extended across the entire planet. Facebook is the world's largest such network. It is the rare high school and college student who does not routinely use Facebook or MySpace. These systems have become so pervasive for communication that young people barely use email anymore. From sixdegrees to Friendster to Facebook, social networking has become a familiar and ubiquitous part of the Internet.

4

Fall 2004 "Look at the world around you. With the slightest push—in just the right place—it can be tipped."

As the fall semester of 2004 loomed, Thefacebook was on the verge of a serious crisis. Over the summer, membership had almost doubled from about 100,000 to 200,000. That was good and bad. "We were just lucky it wasn't completely bringing down the architecture," says Dustin Moskovitz, who spent as much time as anyone working to keep that from happening. "Our servers were already stretched, but we knew a full load in the fall would be double that. Service got really unstable."

But the crisis wasn't just a technological one. Tension was growing among the company's small team about whether Thefacebook itself ought to be their only priority. Zuckerberg was getting increasingly interested in Wirehog, his parallel project to enable Thefacebook users to share photos and other media peer-to-peer.

Throughout the summer, students and student governments from colleges around the country had been emailing, texting, and calling Thefacebook with pleas to add their school to its roster. Sometimes they sent physical letters and included candy or flowers, or even came to the house in Palo Alto. People were literally begging to get into the service.

Saverin was still sitting on the bank account. Zuckerberg was paying bills out of his own pocket. He and his parents had loaned the company many tens of thousands of dollars. But the boys of Thefacebook knew that if they didn't have enough servers when school opened the service might just grind to a halt. "We were really worried we would be another Friendster," recalls Dustin Moskovitz. "We felt that the only reason Friendster wasn't the dominant college network was because they were still having problems scaling." (That's "growing," in non-Net parlance.)

And another "Friendster" was unfolding before their eyes. Orkut, after briefly seeming to rival MySpace, was now getting bogged down with performance problems, in addition to being usurped by Brazilians. Even the great Google couldn't help a social network grow smoothly.

Thefacebook was an atypical start-up, in financial terms. It hadn't required outside funding up to this point. By now—with growth likely and costs rising—a new Silicon Valley company would typically solicit venture capitalists to make a big cash infusion, possibly several million dollars for a company of Thefacebook's size. But in such a scenario, VCs take their pound of flesh—a very big chunk of the company, perhaps a quarter or even a third. Parker had gone through that at Plaxo, where he'd lost the battle of wills with the VCs and been kicked out of his own company. He had successfully infected Zuckerberg with his aversion to VCs. The two were resolute about retaining full control of the company's destiny. After all, they just wanted a few hundred thousand dollars to buy more servers.

Within days after he joined Thefacebook, Sean Parker called his friend Reid Hoffman, the founder of LinkedIn and a big angel investor. Hoffman had been coaching him through the painful denouement of his relationship with Plaxo and had become a close friend. Parker also was pragmatic. He knew it was important to keep that sixdegrees patent close to Thefacebook's camp.

Hoffman was impressed with Thefacebook almost immediately but didn't want to be its lead investor, given his involvement in LinkedIn. By mid-2004, many in the Internet industry were beginning to wonder if social networking might ultimately all converge in one big network. Although Hoffman didn't believe that himself, he knew that some would see it as a conflict of interest if he led an investment in Thefacebook. So he arranged for Parker and Zuckerberg to meet with Peter Thiel, the brooding, dark-haired financial genius who had co-founded and led PayPal and was now a private investor.

Hoffman is a key member of a very distinct and important Silicon Valley subculture—the wealthy former employees of PayPal. Hoffman remained close to many onetime PayPal colleagues, including Thiel. PayPal created the first successful large-scale online payment system and sold itself to eBay for $1.5 billion in October 2002, just two years after it was created in a merger of two start-ups. Even before PayPal, Thiel had been a professional investor and he was now investing in start-ups and starting a hedge fund. He had put money into Friendster as well as LinkedIn.

Thiel turned out to be the perfect investor for Thefacebook. He had seen the world from the perspective of a successful entrepreneur at PayPal. He was a fan of Sean Parker, whom he'd gotten to know at Plaxo and through Friendster. He was also a contrarian thinker. Investors were still generally leery about consumer Internet companies, recalling how much they'd lost when the dot-com bubble burst. "So we felt this was the place to look for opportunity," recalls Thiel. "And within the consumer Internet, social networking seemed to be sort of an incipient trend. But in 2004 social networks were perceived as businesses that were very ephemeral, and people thought it was like investing in a brand of jeans or something. There was a question whether all these companies were just fashions that would only last a very short period of time."

But the message Thiel heard about Thefacebook gave him confidence. In his office sat Sean Parker, Mark Zuckerberg, and Steve Venuto, the company's new lawyer, who had worked on Plaxo with Parker and had started working with Facebook in the late summer. Hoffman has set up the meeting and included his LinkedIn protégé Matt Cohler, an upbeat, brown-haired Yale graduate. Parker, still only twenty-four but already a practiced pitchman, did most of the talking. He explained that while Thefacebook was still relatively small, that was because it required a .edu email address. The potential universe of members was deliberately circumscribed. Only students at select schools could join. What happened once it opened at a new school was what most impressed Thiel. Within days it typically captured essentially the entire student body, and more than 80 percent of users returned to the site daily! Nobody had ever heard of such an extraordinary combination of growth and usage in a start-up Internet company.

Zuckerberg wore his then-standard uniform of T-shirt, jeans, and Adidas open-toed rubber flip-flops. It certainly wasn't calculated to impress. Thiel recalls that he seemed kind of introverted. Zuckerberg said little, occasionally answering questions, and asking a few. Yes, they were receiving requests from hundreds of schools that wanted Thefacebook on their campuses. He talked about some of his ideas for how the product might evolve. He also spoke briefly about his hopes for Wirehog. He showed absolutely no impulse to kowtow, and that, combined with his matter-of-factness about what Thefacebook was achieving, made him seem even more impressive. He didn't need to wear a tie to convince someone he was an entrepreneur worth backing.

But Zuckerberg was also unself-conscious about admitting what he didn't know. The conversation quickly turned to the mechanics of an investment, and Thiel was slinging around technical phrases and jargon about the deal. Zuckerberg repeatedly interrupted: "Explain that to me. What does it mean?"

Within days, following a little back-and-forth with Parker, Thiel agreed to what may go down in history as one of the great investments of all time. He agreed to loan Thefacebook $500,000, which was intended to eventually convert into a 10.2 percent stake in the company. That would give Thefacebook a valuation of $4.9 million. One reason Thiel agreed to provide the money as a loan was because until everything was settled with Saverin there were legal obstacles to a formal stock investment. The provisions of the loan were that if Thefacebook reached 1.5 million users before December 31, 2004—less than six months later—it would convert to an equity investment, and the company wouldn't have to pay it back. Zuckerberg and Parker had a big incentive to keep their growth going.

The $4.9 million valuation was lower than others that had been dangled in front of Zuckerberg, but he was pleased to have found an investor who seemed to believe in giving the entrepreneur the benefit of the doubt. Thiel told Zuckerberg "just don't fuck it up," which the CEO now says was pretty much the only advice he got from Thiel in the company's early years. "I was comfortable with them pursuing their original vison," says Thiel, looking back. "And it was a very reasonable valuation. I thought it was going to be a pretty safe investment." Though Thefacebook didn't meet the December 31 deadline for 1.5 million users, Thiel let the loan convert shortly afterward anyway. Thiel sold almost half his stock in 2009, but even so his remaining shares today are worth—at a minimum—several hundred million dollars. When he made the loan Thiel also joined the company's board of directors.

Hoffman put in an additional $40,000, as did his friend Mark Pincus, and a couple friends of the company invested small amounts, bringing the total financing to around $600,000. Attending the investment presentation also made a deep impression on Hoffman's LinkedIn employee Matt Cohler. He wanted to buy stock in Thefacebook, but Zuckerberg and Parker felt they had enough money for now. Later, however, Cohler would find a way to get some.

Harvard's incoming freshman class in the fall of 2004 was both flattered and shocked when college president Lawrence Summers, former Treasury secretary of the United States, greeted them by announcing he had already acquainted himself with many of them by viewing profiles on Thefacebook. The service was already so entrenched in Harvard's undergraduate culture that incoming students heard about it and created profiles even before they arrived. But the information they had included was intended to impress their peers, not be seen by the president of the college. It made some uncomfortable to think that their personal details and trivia were now so accessible to authorities like Summers. Already Thefacebook was making people wonder how much online self-disclosure was appropriate.

Thefacebook, however, did meet a concrete need among students at Harvard and other colleges. Paper facebooks were handed out freshman year at most schools and typically showed photos of every student along with just their name and high school. Yet for all their limitations they had come to play an outsize role in the social life of schools. If you met a guy at a party, the next morning you'd pull out the facebook to show your roommates what he looked like. If you were by now a junior, the picture would be more than two years old, but it was still the best people had. It was referred to at some schools as the "Freshmenu," and people would play games with it if they were bored on Friday nights. Open to a random page. The winner was the one who had to turn the fewest pages before they came to ten people they'd slept with. Things like that.

So at Harvard, Dartmouth, Columbia, Stanford, Yale, and other schools, Thefacebook quickly became an essential social tool—a considerable advance over the outdated paper book. Now if a girl met a guy at a frat party, an elaborate set of electronic rituals was set in motion. They took on even more significance if you had already "hooked up" (slept together). The first key question was whether the guy immediately friended you on Thefacebook. If he didn't, that was a disastrous sign. At the time, any student could see everyone else's profile at his or her school. Another urgent key activity was to closely examine the friends of your new acquaintance. Thefacebook told you which friends you shared. A large number was a promising indicator.

Thefacebook had a strong sexual undertone. You were asked to list your relationship status and whether you were interested in men or women. One of the site's standard data fields was labeled "Looking for." Possible answers included Dating, A Relationship, Random Play, and Whatever I Can Get. Flirting on Thefacebook became a sort of art form, though one feature—the poke—made doing so absurdly easy.

Poking was a particular fascination in those days, even among the supposedly sophisticated students of Harvard. There was no certainty that a poke on Thefacebook would be seen as flirting—at least in theory it could be construed as just a friendly gesture—so even the shy occasionally found the gumption to click on it. The very fact that the meaning of a poke was so indeterminate was one of its appeals. It could mean you liked someone, found them attractive, enjoyed their comment in class, wanted to distract them from their homework, or just wanted their attention. The recipient was only told that he or she had been poked, so was left to interpret that information however they would. The proper response? A poke back, which Thefacebook's software politely inquired if you'd like to do.

"Friending" had an element of competitiveness from day one, as it had on Friendster and MySpace. If your roommate had 300 friends and you only had 100, you resolved to do better. "Competition definitely caused Thefacebook to spread faster at Dartmouth," says Susan Gordon, class of 2006. She was on an Italian study program in Rome when Thefacebook blanketed Dartmouth almost overnight in March 2004. She immediately began receiving emails from friends telling her that she had to join, otherwise she would be way behind when she returned at the end of the quarter. "It made a lot of sense to all of us immediately," she says. "An online green book—how fun!" (Dartmouth's facebook had a green cover.) The fact that it was Ivy League–only was also reassuringly exclusive. It had started at Harvard. It must be okay.

Perfecting the details of your own profile in order to make yourself a more attractive potential friend occupied a considerable amount of time for many of these newly networked Ivy Leaguers. Find exactly the right picture. Change it regularly. Consider carefully how you describe your interests. Since everyone's classes were listed, some students even began selecting what they studied in order to project a certain image of themselves. And many definitely selected classes based on who Thefacebook indicated would be joining them there. A subtle form of stalking became almost routine—if someone looked interesting, you set out to get to know them. The more friends you already shared the easier that process would usually turn out to be. Your "facebook," as profiles on the service began to be called, increasingly became your public face. It defined your identity.

People spent hours and hours visiting the profiles of other students, initially just at their own school but soon across Thefacebook's network at other elite campuses as well. Nick Summers, Columbia '05, who was user number 796 on Thefacebook, recalls being able to browse through the faces of every user on the entire service from A to Z. There wasn't that much you could do there except maintain your own profile, add friends, poke people, and view the profiles of others. Nonetheless, students spent thousands of hours examining every nook and cranny of others' profiles. You could ask Thefacebook to display ten random students from your school for you to peruse, or you could search for people based on various parameters. The latent nosiness and prurience of an entire generation had been engaged.

In September, Thefacebook added two features that gave students even more reason to spend time there. Now included on user profiles was something called "the wall," which allowed anyone to write whatever they wanted right on your profile. It could be a message to you or a comment about you—the equivalent of a public email. Any visitor to your profile could see it. Now not only could you surf around examining people, but you could react to what you learned. Or you could simply invite someone to meet you in the cafeteria later. Or make a seductive comment. And another friend could comment on that comment in his or her own post to the wall. Suddenly every Thefacebook user had their own public bulletin board.

Over the summer, Zuckerberg, Moskovitz, and Parker had coined a term for how students seemed to use the site. They called it "the trance." Once you started combing through Thefacebook it was very easy to just keep going. "It was hypnotic," says Parker. "You'd just keep clicking and clicking and clicking from profile to profile, viewing the data." The wall was intended to keep users even more transfixed by giving them more to see inside the service. It seemed to work. Almost immediately the wall became Thefacebook's most popular feature.

The other new addition was Groups. Now any user could create a group on Thefacebook for any reason. Each group had its own page, much like a profile, which included its own wall-like comment board. Instantly, inane groups sprouted at Harvard with names like "I puke Vitamin Water," which for some reason quickly garnered 1,000 members. Emma MacKinnon, Harvard '05, was writing her senior thesis on the philosopher Emmanuel Levinas and remembers belonging to a group on Thefacebook called "I hate the guy my thesis is about." Its members were women writing about men. "There was always also a little description about 'Why I really love him,'" she recalls.

Many students began to abandon their address books because they could use Thefacebook to contact anyone by simply entering their name. You didn't need to remember or store anyone's email address. And if you wanted to reach someone immediately, almost everyone listed both their cell-phone number and their AIM instant-message address in their profile. These were not the anonymous identities of an AOL chat room. The Internet had entered a different era. It was becoming personal.

Thefacebook's full-time California staff was down to just Zuckerberg, Moskovitz, Parker, and operations manager Halicioglu, who stayed in San Jose, twenty miles south. Andrew McCollum lived with them, too, still focusing on Wirehog. Thiel's money enabled them to buy new servers, and they were expanding furiously. In the first week of the fall semester they added fifteen or so new colleges. Thefacebook was quickly losing its elitist edge. By September 10, the list included places like the University of Oklahoma and Michigan Tech.

Since they had to move out of the trashed summer sublet, they found a new rental in Los Altos Hills, a few miles south. Its yard backed onto Interstate 280. That helped solve problems with neighbors—it was so noisy there all the time that nobody noticed parties and late-night antics. But the construction dust left over in the brand new house made the house almost uninhabitable for Sean Parker. His allergies went into crisis mode there. Luckily his girlfriend let him stay at her place.

They were good at running a website, but not so good at running a house. They turned the living room into a makeshift office with whiteboards on the periphery, tables scattered with laptops, and papers strewn everywhere. One of Zuckerberg's high school friends stopped by and found that all the tables had been pushed to one side of the room. They'd overloaded the wiring and tripped a circuit breaker, which shut off power to some of the outlets. Rather than search for the circuit box, they'd just moved over to the remaining outlets. The visitor found the circuit breakers and flipped a switch, so the geeks could spread back out. Most of the rest of the house was empty, except for mattresses here and there and a bunch of never-unpacked boxes.

Hygiene was, if anything, deteriorating. In the kitchen, dirty dishes spilled out of the sink and nobody ever emptied the trash. Ants were everywhere. Zuckerberg continued leaving empty drink cans wherever he happened to finish them. This is how twenty-year-old college kids live.

While the housekeeping may have been immature, the company increasingly wasn't. With its new structure and rapidly growing membership, Thefacebook seemed to be growing up. Demand was even more ferocious than they'd expected. In September alone they nearly doubled membership, to around 400,000. The number hit half a million on October 21, as growth began to accelerate. They'd figured out over the summer how to automate much of the process of adding a new school, so the painstaking assembly of dorm lists and class schedules was gone.

But it was quickly becoming clear that even Thiel's money wasn't enough to cover all the costs of the rapidly growing company's infrastructure. Adding new servers was an almost daily activity. Sean Parker, whom Zuckerberg had deputized to handle all financial matters, got in touch with a firm called Western Technology Investment, which he knew from his Plaxo days. WTI, as it is known, is in the business of "venture lending." It makes short-term loans—usually repayable within about three years—to start-ups at interest rates ranging from 10 percent to 13 percent. Maurice Werdegar, a partner at WTI, negotiated with Parker a $300,000 three-year credit line that was specifically allocated to cover Thefacebook's costs of computer hardware and other physical assets, which WTI put a lien on until the loan was repaid. The loan closed in December 2004 with its credit line expected to run out by the following July. Werdegar, who was a big fan of Parker's and had not had any difficulty dealing with him at Plaxo, liked what he was hearing from him about Thefacebook's prospects. He asked if WTI could in addition invest $25,000 at the same company valuation Thiel had gotten.

Werdegar, at the time a junior partner in the firm, arranged for the two founders of WTI to join him for a meeting in October 2004 with Parker and an accountant named Mairtini Ni Dhomhnaill (she was Irish) who was temporarily working for Thefacebook. Before the meeting, Werdegar's partners told him that while they were happy to be dealing with Sean Parker again, this company didn't seem to merit a stock investment on top of the loan.

That attitude quickly changed. "After an hour and a half listening to Sean we all walked out and—I will forever remember this—they asked

'How much of that equity can we get?'" recalls Werdegar. Parker was refining his rap, and the facts were just getting better and better.

The company's latest governance structure, following Thiel's investment, had some unusual provisions. The board of directors included four seats: One was held by investor Thiel, one by Parker, and one by Zuckerberg. The fourth seat was Zuckerberg's to allocate as he saw fit, and remained for the time being vacant. The idea was to outnumber outsiders and make it impossible for any future investor to usurp the company. Since Thiel was himself a former entrepreneur and a believer in letting founders control their creations, it didn't bother him.

This is a key way Parker put his stamp on the company. He had been fired twice—from Napster and from Plaxo. He didn't want to be fired from Thefacebook, and he wanted to make it impossible for Zuckerberg to get fired, either. "What I told Mark," says Parker, "was that I would try to be for him what no one had been for me—a person who sort of shepherds his rear and puts him in a position of power so he'd have the opportunity to make his own mistakes and learn from them." Another provision in the company's corporate documents guaranteed that if any of the founders, or Parker, ever left the company for any reason, they would be able to keep both their company email address and their company laptop. After being ousted from Plaxo, Parker had found himself without either, and thus nobody could get in touch with him.

"It was really beneficial to us that Sean had been a founder who had been burned," says Moskovitz. "We didn't know anything about how to incorporate a company or to take financing, but we had one of the most conservative people figuring it out for us and trying to protect us." When he calls Parker conservative he is referring not to his personal style. Parker could be erratic. But Moskovitz recalls his role at the company in those days fondly, despite some unhappy incidents later. Parker was old enough to buy the alcohol, he knew how to throw a party that involved more than Beirut beer pong, and he could talk the language of venture capital. "It was just a comfort to have him around," says Moskovitz. "Sean is like excitable and kind of a crazy person, but it just generally makes life more interesting when you're a bunch of geeks sitting around a table hacking all day." Parker, twenty-four, seemed a sophisticate to these twenty-year-olds.

Zuckerberg may have been more focused and steady than Parker, but he too had his quirks. They were, however, mostly intellectual and verbal. For instance, he had a way of punctuating a conversation, when it reached a critical moment, by suddenly pronouncing, "Now you know who you're fighting!" It was a quote from one of his favorite movies, *Troy*, which he had seen on Harvard Square with friends on his twentieth birthday the previous May. Zuckerberg had loved studying the classics. In a key scene from the movie, the Greek warrior Achilles, played by Brad Pitt, confronts his Trojan adversary Hector:

> *Hector*:   Let us pledge that the winner will allow the loser all the proper funeral rituals.
> *Achilles*:   There are no pacts between lions and men.
> [stabs spear into ground, and takes off helmet, throwing it to the side]
> *Achilles*:   Now you know who you're fighting.

As a fencer—the civilized version of a swordsman—Zuckerberg sometimes saw the world as a fencing match, a forum for combat in which the ideal thrust is the one that catches an opponent off guard.

One battle that had become engaged was the one with the Winkelvoss/Narendra axis. The company had by now hired another law firm just to defend itself against the lawsuit, which was beginning to wend its way through federal court in Boston and had become a significant magnet for the company's limited assets, costing about $20,000 per month in lawyers' fees. After a phone call from the lawyers, Zuckerberg would briefly recount the latest news to Moskovitz, then stand up and declaim, "Now you know who you're fighting!" At other moments, the phrase made even less sense.

But incongruous movie quotes gave Zuckerberg, who could otherwise frequently lapse into long periods of silence, tremendous if inexplicable pleasure. He also inserted them inside Thefacebook. Whenever you searched for something in those days there was a little box below the results. Initially it had some tiny type that said, "I'll find something to put here." Later that was replaced with "I don't even know what a quail looks like." It's a throwaway line from *The Wedding Crashers*. Another quote that appeared there was "Too close for missiles. Switching to guns," which is spoken at a critical moment in *Top Gun* by the fighter pilot played by Tom Cruise.

The quotes came to encapsulate, in the fashion of schoolboy in-jokes, the spirit of the company—playful, combative, and despite the technical sophistication, a bit juvenile. Students at colleges around the United States spent hours arguing about the significance of these inscrutable epigrams. Not long afterward, Aaron Sittig designed company T-shirts. They showed a fighter plane streaking past a couple of quails.

The twelve-year-old Ford Explorer they'd purchased over the summer finally gave out. One day it just wouldn't start, with or without the key. Zuckerberg and Parker brought it up at their next board meeting with Thiel, who approved buying a company car. "Just keep it under fifty," he exhorted them. They bought a sleek black new Infiniti FX35, a luxury sport utility vehicle. The car has a streamlined, vaguely malevolent look, as if poised on its haunches to leap on top of some unsuspecting Ford. Now you know who you're fighting! They nicknamed it "the Warthog," after a tank in the video game Halo, which they played regularly on their Xbox. The toys were getting better.

Facebook seemed to be thriving, but Zuckerberg was thinking about Wirehog almost as much. "What was so bizarre about the way Facebook was unfolding at that point," says Sean Parker, "is that Mark just didn't totally believe in it and wanted to go and do all these other things." Zuckerberg felt he had cause to hedge his bets. He was worried that once Thefacebook began trying to expand beyond college it would hit a wall of resistance. He was genuinely unsure which of his projects would ultimately lead to the best business. And it wasn't just about business. Zuckerberg hadn't changed much since he was just out of Exeter and turned down millions for Synapse, the MP3-playing tool he built with D'Angelo. The ideas were at least as interesting as the prospect of riches.

In any case he was confident one of his projects would really catch fire. Maybe it would be Wirehog. "Mark always talked about how he just liked to start things, especially back then," says Dustin Moskovitz. "He was like 'My life plan basically is I'll prototype a bunch of these apps and then like try and get people to run them for me.'"

Parker, on the other hand, remained resolutely opposed to Wirehog. "I specifically said Wirehog is a terrible idea and a huge distraction. We shouldn't be working on it,'" he recalls. Yet Zuckerberg prevailed upon Parker, who reluctantly hired Steve Venuto, the same lawyer who had set up Plaxo, to create Wirehog the company. In order to try to seduce Parker's support, Zuckerberg made him one of Wirehog's five share-holders, along with McCollum, D'Angelo, and Moskovitz, who himself was ambivalent about Wirehog. "I needed Mark's attention on Face-book," remembers Moskovitz. Looking back, Zuckerberg concedes that he didn't always make things easy for his partners: "Dustin was fully bought in to what we were doing [with Thefacebook]. And I was always thinking about the next thing. Until we got to this big inflection point, for my calculus it may have just not been worth the amount of work."

The group was split. McCollum and D'Angelo focused almost en-tirely on Wirehog, while Parker and Moskovitz worked only on The-facebook. Zuckerberg straddled both projects. "Wirehog was more interesting to a lot of us," says D'Angelo. "There were a lot of social networks. Wirehog was a product I was interested in using myself, and it was more technically interesting, too."

Wirehog was a stand-alone program that users downloaded onto their computers. The entrepreneurs built a little profile box for it on Thefacebook, so it could figure out who your friends were and whether they too had downloaded Wirehog. It gave you a window into their computer and told you what files they were willing to share. Wirehog was intended primarily for photos, since that was what Thefacebook's users were most vociferously asking to share with one another. (At the time you were only allowed one photo of yourself on your profile page.) But Wirehog could also handle video, music, and documents. "We kind of thought of Wirehog as the first application that was built on top of Facebook," says D'Angelo. Zuckerberg also talks about Wirehog as the first example of treating Thefacebook as a platform for other types of applications. D'Angelo kept writing code for Wirehog in the fall after he returned to Caltech.

Over the protestations of Parker and Moskovitz, Wirehog launched in November 2004 as an invitation-only site at a few colleges. A page on Thefacebook explained: "Wirehog is a social application that lets friends exchange files of any type with each other over the web. Thefacebook and Wirehog are integrated so that Wirehog knows who your friends are in order to make sure that only people in your network can see your files." Its site listed things you could do with it: "share pictures and other media with friends; browse and save files through the web; roll around in the mud oinking and such; transfer files through firewalls."

But Wirehog was too complicated for most users of Thefacebook, just as Parker had predicted. He was desperate to shut it down, so Thefacebook could avoid the prospect of a debilitating lawsuit. They may have been separate companies, but Thefacebook was where a user went to download the Wirehog software. Soon even Zuckerberg began to cool on Wirehog. "He just like came back to reality on how much time he was spending," says Moskovitz.

On November 2, MySpace hit 5 million users. The boys in Los Altos Hills were bemused. They saw themselves creating an anti-MySpace. Where that service was wide-open, florid and unconstrained, Thefacebook was minimal, with limited flexibility and no decorative freedom. MySpace was unconcerned with who you really were. Thefacebook authenticated you with your university email, and you had no choice but to identify yourself accurately. On MySpace, the default setting was that you could see anybody's profile. On Thefacebook, the default allowed you only to see profiles of others at your school, or those who had explicitly accepted you as a friend. A degree of privacy was built in. "On MySpace people got to do whatever they wanted on their profiles," says Zuckerberg. "We always thought people would share more if we didn't let them do whatever they wanted, because it gave them some order."

While Zuckerberg was relatively unflustered by MySpace's advances, he worried much more about collegiate competition. A variety of other university-centric social networks was quickly rising. It became a top company priority to stamp them out. One, called CollegeFacebook.com, was a pure copycat, in look and tone. Its strategy was to go after less snooty schools that the elite Thefacebook hadn't yet targeted. It briefly garnered hundreds of thousands of users until the real thing arrived. The Winkelvoss/Narendra team had finally launched Harvard Connection in May, now dubbed ConnectU to emphasize it was for any school. Columbia's CUCommunity, similarly renamed Campus Network, was also expanding to other campuses and becoming established. These competitors were expanding more quickly than Thefacebook. They typically had fewer users at each new campus so they could add more users without putting as much demand on their systems as Thefacebook's hordes did. But their widening presence still made Zuckerberg and his partners nervous.

So they embarked on what they called a "surround strategy." If another social network had begun to take root at a certain school, Thefacebook would open not only there but at as many other campuses as possible in the immediate vicinity. The idea was that students at nearby schools would create a cross-network pressure, leading students at the original school to prefer Thefacebook. For example, Baylor University in Waco, Texas, had one of the earliest homegrown college social networks. Thefacebook launched at the University of Texas at Arlington to the north, Southwestern University to the southwest, and Texas A&M University to the southeast. This pincer movement tended to work, since Thefacebook typically saw such a viral explosion when it launched at schools that didn't already have a social network on campus. Zuckerberg was only twenty, but already he was strategically outmaneuvering his competitors.

Zuckerberg took a utilitarian approach to advertising. If costs were going to go up, ad revenues needed to as well. He wanted to make sure Thefacebook earned enough to cover its costs, which were hovering at around $50,000 a month. "If we're gonna need $100,000 worth of servers or $500,000 worth of new people . . . then how much advertising do we need now?" he asked rhetorically at the time, in an interview with the *Harvard Crimson*.

Y2M, the advertising representative for the company, scored a coup in August with a breakthrough ad deal from Paramount Pictures to promote the November premiere of *The SpongeBob SquarePants Movie*. In those days the only ads on Thefacebook were long vertical rectangles on the lower left side of a page. Paramount paid Thefacebook $15,000 for 5 million impressions (what ad people call a $3 CPM, or cost per thousand views). Paramount also pioneered a concept that would later become a critical piece of Thefacebook's commercial structure—a special group for fans of the movie. The ad urged users to join the group, which consisted mostly of a discussion board. The "group description" read, "Everyone sing it with me now! 'Who lives in a Pineapple under the sea?'" Some users thought the whole thing idiotic and said so on the movie's page. There were as many comments insulting people who liked the movie as there were notes from fans. Nonetheless, the experiment was deemed a success. Over 2,500 users of Thefacebook mentioned the movie in their profile.

By December, Y2M had signed a landmark deal with Apple Computer. Not only did Apple sponsor a group on Thefacebook for fans of its products, but it also paid $1 per month for every user who joined, with a monthly minimum of $50,000. The group was immediately popular, and the minimum was easily exceeded. This was by far the biggest financial development in Thefacebook's short history. It alone more or less covered the company's expenses. Executives at Apple were thrilled because they had acquired a powerful platform which allowed them to be in constant touch with Apple fans in college, whom they began offering special discounts and promotions like free iTunes songs. The deal gratified Zuckerberg because it wasn't a conventional banner advertisement, which he detested.

There were also more modest ads students could buy themselves right on the site, called flyers. A flyer could be targeted just for students at your school. At even the largest campus the ads cost less than $100 per day. It was an effective way for campus groups to promote activities or for a fraternity to announce a big party. The boys wanted to take this further, to create a system so merchants in college towns could buy ads that targeted students. So Parker recruited a new employee, a former roommate of his named Ezra Callahan, who had sold ads for the *Stanford Daily* when he was a student there. Parker offered him the job by email while Callahan was traveling in Europe. A few weeks later he showed up direct from the airport at 1 A.M. Parker was at a movie with his girlfriend and hadn't told the others much about the new hire. So when Moskovitz blearily opened the door he had no idea who Callahan was. "I'm Ezra. I work for you guys," insisted Callahan. Moskovitz let him in. Callahan got a bunch of stock options, which he assumed would be worthless. He planned on going to law school soon anyway. Though the local business product occupied much of his time and that of many others in ensuing months, it never launched. Callahan's other job was to learn from Saverin how to manage and schedule all the other ads. The "CFO" had been doing it remotely from the East Coast. Even though Zuckerberg had for now reached a détente with his erstwhile partner, he was whittling away Saverin's remaining responsibilities.

On November 30, Thefacebook registered its millionth user. It had existed for just ten months. Peter Thiel had recently opened a nightclub and restaurant in San Francisco called Frisson. He offered up its VIP lounge for a party. Parker, the organizer, piggybacked on it a celebration for his own twenty-fifth birthday on December 3.

The email invitations were portentous. At top was a quote from Malcolm Gladwell's *The Tipping Point*: "Look at the world around you. With the slightest push — in just the right place — it can be tipped." The invitation continued: "And tip it has . . ." It was indicative of the way Parker talked about the company. In his opinion the company's success was essentially a fait accompli. Even Zuckerberg was amazed they had gotten to a million so soon.

D'Angelo and Saverin flew in, along with public relations manager and former Zuckerberg roommate Chris Hughes. Investors, friends, and hangers-on drank the night away in glitzy surroundings, regardless of the fact that several of the core team were not yet of legal drinking age. And there were still questions about where priorities lay. One guest at the party asked D'Angelo what he did at Thefacebook. "Oh no," D'Angelo answered. "I work at Wirehog."

Facebook's success was beginning to make waves. And in Silicon Valley, success attracts money. More and more investors were calling. Zuckerberg was uninterested.

One of the supplicants was Sequoia Capital. Among the bluest of blue chip VCs, Sequoia had funded a string of giants—Apple, Cisco, Google, Oracle, PayPal, Yahoo, and YouTube, among many others. The firm is known in the Valley for a certain humorlessness and a willingness to play hardball. Sequoia eminence grise and consummate power player Michael Moritz had been on Plaxo's board and was well acquainted with Sean Parker. It was not a mutual admiration society. Parker saw Moritz as having contributed to his downfall. "There was no way we were ever going to take money from Sequoia, given what they'd done to me," says Parker.

But it seemed a good idea at the time to pitch them instead on Wirehog, as a joke. It was ludicrous, but aside from sticking it to Sequoia it had another, more symbolic purpose. Zuckerberg's acquiescence seems to have been his way to surrender to Parker about Wirehog—to acknowledge that it was dead. "We had done a couple of real pitches for Wirehog, but our theory was that no one cared about that," says Zuckerberg. "They just wanted to do Facebook." Sequoia, for its part, was so eager to get close to them that partner Roelof Botha willingly accepted the idea.

The boys hatched a plan. On the appointed day, they overslept. They were supposed to meet at 8 A.M. Botha called at 8:05—"Where are you guys?" Zuckerberg and Andrew McCollum, his Wirehog partner, rushed over to Sequoia's swanky offices on Menlo Park's Sand Hill Road in pajama bottoms and T-shirts. Though they said they'd overslept, it was deliberate. "It was actually supposed to be worse," says Zuckerberg. "We won't even go there." Then, as the stiff but attentive partners of Sequoia looked on, Zuckerberg made his presentation.

He showed ten slides. He didn't even make a pitch for Wirehog. It was a David Letterman–style list of "The Top Ten Reasons You Should

Not Invest in Wirehog." It started out almost seriously. "The number 10 reason not to invest in Wirehog: we have no revenue." Number 9: "We will probably get sued by the music industry." By the final few points it was unashamedly rude. Number 3: "We showed up at your office late in our pajamas." Number 2: "Because Sean Parker is involved." And the number one reason Sequoia should not invest in Wirehog: "We're only here because Roelof told us to come."

Throughout it all, the partners seemed to listen respectfully, recalls Zuckerberg, who says he now deeply regrets the incident. "I assume we really offended them and now I feel really bad about that," he says, "because they're serious people trying to do good stuff, and we wasted their time. It's not a story I'm very proud of." Sequoia never invested in Facebook.

Now, at last, Thefacebook was the sole priority. Not only the world, but Zuckerberg himself had been tipped. President Parker started looking for top talent to fill out its staff. One of the first people he targeted for a senior position was Matt Cohler, Reid Hoffman's right-hand man, who had attended the Thiel investment pitch. Cohler had a combination of qualities Parker liked. He was a natural intellectual, well-versed in the exigencies of the Internet, with extreme social dexterity. Cohler had a degree in musicology with honors from Yale, so he fit in nicely with Thefacebook's Harvard crowd. And he even had international experience, having lived in China while working for an Internet company there. But he was pretty happy at LinkedIn, which at the time was seen as one of the hottest and most promising of the Silicon Valley start-ups.

Cohler talked to a bunch of friends trying to figure out if he should really consider Parker's offer. He was almost twenty-eight—no longer a college kid—and had even spent time at the venerable McKinsey consulting firm. He wasn't one to make rash moves. Cohler called his brother, an undergraduate at Princeton, to ask if he knew about this thing Thefacebook. "The answer was like, 'Duh!' as if I'd asked 'Do you guys have electricity at Princeton?'" remembers Cohler. But he found the numbers Thefacebook was claiming hard to believe.

He asked Zuckerberg if he could spend some time poking through the service's database on his own. Cohler was blown away by what he found. He, Parker, and Zuckerberg came to an agreement shortly afterward. At that point everybody in the company was earning $65,000 a year, plus—and this was critical to Cohler—a fair amount of stock. Cohler was convinced Thefacebook could get big. He had no interest in Wirehog. His job was to help Thefacebook become a real company. His role, as he later put it, was to be Zuckerberg's "consigliere."

5

Investors "I've got to invest in this company."

One of Chris Hughes's friends at Harvard's Kirkland House was Olivia Ma, whose father, Chris, was a senior manager for acquisitions and investments at the Washington Post Company. Ma's daughter urged him to take a look at Thefacebook, and between Christmas and New Year's of 2004 he took Zuckerberg to a Sunday lunch in Menlo Park, near Facebook's offices in Palo Alto.

The Post was already an investor in Tribe.net, and Ma found Thefacebook enticing because of its focus on a promising demographic—college students. He also immediately found himself impressed with Zuckerberg. "I concluded in that first lunch that the key to Mark is that he is a psychologist," says Ma. "His central thought was that kids have a deep-seated desire to have certain kinds of social interactions in college and that what drives them is their extreme interest in their friends—what they are doing, what they are thinking, and where they are going. He had some simple but deep insights." Ma talked a little about the Post's hands-off aproach to its investments and suggested that the company would be interested in putting money into Thefacebook. Zuckerberg said he would think about it. He and Ma formed a bit of a bond not only because of Olivia, but because Chris had attended Exeter Academy, Zuckerberg's prep school. Two weeks later Zuckerberg called and told Ma he wanted to come to Washington to discuss the possibility of an investment.

Sean Parker, whose family lived near Washington, joined Zuckerberg for the trip, and when they arrived at the Post, they found not only Ma and another investment executive, but Post CEO Don Graham crowded into a tiny conference room. The Post's offices were in the midst of a renovation. Ma had asked Graham to join them briefly to meet this impressive young entrepreneur. The ruddy-faced Graham is a renowned leader of American business and a member of the family that has controlled the *Washington Post* since the 1930s. Zuckerberg explained Thefacebook to Graham, who recalls being immediately taken. "I thought it was a simply stunning business idea," he says. Hearing about Thefacebook's success at Harvard elicited an oddly relevant recollection in Graham, then fifty-nine. He too was a Harvard man and was taken back to his days as a reporter and president of the *Harvard Crimson* in the mid-1960s.

In those days the *Crimson* kept several large ruled ledger books on a shelf in its newsroom. Articles from each day's paper were pasted into the books, and staffers wrote comments about them right there. In another book they wrote anything that was on their minds. "I vividly remembered that every time any of us would walk into that room we would read every word written in those ledgers and write our own comments," says Graham. "I have often thought about the power of those comment books and wondered whether there was some way to replicate them in a place like this." (At his newspaper, that is.) "When I heard Mark describe the idea of Thefacebook, I thought 'Oh my God, I see exactly what he is shooting for.'" Graham was then deeply immersed in efforts to build the Post's own Web businesses.

He was amazed to hear how many hours users were spending each day on Thefacebook. He was also somewhat amazed by Mark Zuckerberg. "Mark was ever so slightly more awkward than he is now," continues Graham. "If you said something he would pause and think about it before he'd comment or react. But every single thing he said in the course of that conversation made a lot of sense. It was remarkably impressive for a twenty-year-old."

Graham began recounting a bit of the company's history, and Parker remembers him saying something like ". . . and then a man named Warren Buffett came into our lives." Neither Zuckerberg nor Parker knew much about the Graham family before their visit, but they had heard of Buffett, the legendary investor and one of the world's richest men. The Berkshire Hathaway company, run by Buffett, has been a large investor in the Post since the 1970s. "He said the arrival of Buffett was a transformative moment in the life of the company," says Parker. Graham explained that the Post was able to take a very long-term view of its corporate strategy, both because the Graham family controls a huge portion of its voting stock and because Buffett has made clear he intends to hold his shares for the long term.

At some point in their conversation, Graham made an offer more spontaneous than any he says he had ever made before or has made since. He recounts it: "I said, 'Mark, in the end you will not do this, but if you wanted an investor who wasn't a venture capitalist and wouldn't pressure you in any of the ways VCs normally do'—he had talked a little about how he did not want to go that route—'we'd probably be willing to invest.'"

Zuckerberg was deeply impressed with how Graham thought about business. He explains: "A lot of VC firms had approached us, but I didn't want to play this whole Silicon Valley game of—take VC money, try to go public or sell the company really quickly, bring in professional management on an accelerated time scale—things like that. But the Washington Post is a completely different kind of company than these technology companies. I was just blown away by the difference in culture, that it's just such a long-term focus there, and that they're so focused on the brand of the Washington Post and the trust it has. I was just like 'Wow. I want to be more like this guy.' And that's when I seriously started thinking about doing another [investment] round. And I was looking forward to doing it with them. Don was a guy I could work with." Graham stayed ninety minutes, far longer than he had planned. When he finally had to get up to go, Zuckerberg rose as well. The twenty-year-old looked Graham in the eyes. "You're cool," he said. Graham smiled broadly.

Things began moving quickly after that. The Post Company sent another, larger delegation out to Palo Alto. The top managers of the Post's online division joined the trip—including newly named CEO Caroline Little, along with the vice presidents for finance and business development. Little says Thefacebook seemed like a potential gold mine. "Mark was kind of against ads, as far as we could tell," she says. "But I just sat there salivating and thinking how easy it would be to monetize this. I had to bite my tongue because Don didn't want to hear it."

Sean Parker interpreted the turn in Zuckerberg's attitude toward investment as a license to more aggressively beat the bushes and see what else was out there. "I thought we should be valued at half a billion dollars even then," Parker says now, a bit bombastically. "It was pretty obvious to us that we were taking over the world." But in reality, at this point even Parker was talking about a relatively low valuation. His friend Seth Sternberg (now CEO of messaging company Meebo) recalls Parker asking for valuation advice. Though Sternberg was only twenty-six, he had worked in corporate development at IBM, so his opinion counted. Sternberg recommended shooting for a valuation of at least $40 million. So Parker raised his expectations to a target range of $40–60 million, company documents show. Anything like that would be phenomenal for a year-old company led by a twenty-year-old with seven employees and annualized revenues of less than $1 million. But Parker was wrong. He got more.

As soon as word got out that Thefacebook was contemplating an investment, the Silicon Valley greed machine kicked into high gear. Inquiries started pouring in. Cell phones at Thefacebook rang incessantly. Ron Conway, one of the most connected men in Silicon Valley and a veteran angel investor, was giving Parker advice about who to talk to and what to say. He was also making email introductions to established Silicon Valley companies and key venture capital firms. (Zuckerberg now says he didn't know about most of this activity at the time.)

Investor interest was further heightened when the *Los Angeles Times* wrote a front-page story about Thefacebook on January 23, the first big story ever about the company in a major media publication. SITE INTRIGUES COLLEGIANS ACROSS U.S., read a headline. "Clever, goofy, or profane, the website has a powerful hold on its members," Rebecca Trounson wrote, "most of whom log on to it almost every day."

Parker got his friend Sittig, now Thefacebook's design guru, to design a few PowerPoint slides and started meeting with potential investors. The six-page presentation was modest but compelling. It claimed Thefacebook had 2 million active users (this was mid-February) and was deployed at 370 schools. But what got investors' attention was the data about how engaged users were. An amazing 65 percent of them were returning to the site daily, and 90 percent came back at least once a week. Growth was so torrid that it occasionally hit 3 percent per *day*.

What most wowed those who saw the presentation was a simple growth chart. Parker and Sittig had designed it to afford a little drama. At first Parker would put up a chart showing colleges where the service had opened. It looked like a staircase, because Thefacebook opened schools in batches and then didn't add new ones for a while. Then another slide would overlay on top of the school chart. It showed the trend in total users. Vividly apparent was a correlation between opening schools and growth in users. After each step in the staircase of schools, the number of users leapt upward with minimal delay. That implied the possibility of near-guaranteed growth, at least until Thefacebook saturated the available population of 16 million American college students.

The simple business plan Parker presented explicitly avoided mentioning conventional Internet banner ads, even though they had been the primary source of the little income Thefacebook had gotten thus far. Page four of the PowerPoint was titled "Local Advertising." It projected that on-campus text-based "flyers"—used by campus organizations to announce events—would yield net sales of $3.65 million a year if Thefacebook was at 400 schools. But the separate, yet-unlaunched local advertising product—the one Parker had hired Ezra Callahan to spearhead—was projected to be much bigger, yielding annual sales of $36.6 million. That assumed sixty businesses would advertise at each of 400 schools, offering students discounts, coupons, etc.

Then there was a page of sheer marketing chutzpah. The presentation touted what it called "AdSeed," defined as "Google AdSense for social networks." Google's AdSense is a program that gives websites revenue in exchange for allowing the search company to place textual advertising on their pages based on the content there. At the time it was beginning to take off. With the AdSeed name Thefacebook was trying to feed off Google's buzz. The slide explained: "Products, brands, and media properties (movies, books, music) receive a 'home' in social-space." At the time, Thefacebook was just beginning its lucrative arrangement for Apple's sponsored page, and that was the model. The slide claimed Thefacebook was at that time making $40,000 a month from pilot customers for AdSeed. The name, by the way, was never actually used.

By February 9, twelve venture capital firms, four major technology companies, and the Post were actively pursuing Thefacebook for some kind of deal, according to a company document. Parker had decided not even to pursue investors who needed convincing. A few well-known outfits were out of the running. Kleiner Perkins and Benchmark, two of the Valley's most eminent firms, were both already up to their ears in social networks thanks to their troubled investments in Friendster. Neither wanted anything to do with Thefacebook.

For all the impressiveness of the PowerPoint data, there remained significant skepticism toward Thefacebook. After all, the only way any potential investor could even log on and take a look was with an alumni email address from their alma mater. A limited-access consumer website was something new. Then there was the personnel matter—an inexperienced twenty-year-old CEO and a partner with a profligate reputation.

But even as the talks with the Post and others were heating up, Thefacebook needed money right away. Parker decided to borrow some more from his friend Maurice Werdegar at WTI. The initial $300,000 credit line had been used up after less than two months, even though it was expected to last for eight months. Parker wanted to borrow another $300,000. But he and Werdegar couldn't agree on how to value the accompanying warrants that would give WTI the right to purchase Thefacebook's stock. WTI typically only invests following a financing by venture capitalist firms, and the prices of its warrants are typically pegged to what the VCs have already paid. It had stretched its practice somewhat by putting money in alongside an angel investor like Thiel.

Werdegar thought WTI was taking a significant risk, because after this loan its total outlay for Thefacebook would be $625,000, counting both loans and the $25,000 equity investment. He wanted the warrants attached to this new loan to be priced at the same level as the ones for the first one only months earlier—the price that Thiel had paid for Thefacebook's stock.

Parker started talking about how great the company's next fundraising round was likely to be, which meant that WTI would be very likely to be repaid with little difficulty. If that was true, the risk was small. Parker wanted the warrants to be priced based on the upcoming venture capital round. Parker said the valuation was likely to be at least $50 million, which Werdegar found ridiculous. Parker also asserted that an investment was imminent. Werdegar didn't believe him.

So they made a bet. Written into the loan's terms was a provision that if Thefacebook closed a venture round of $2 million or more by May 15, 2005 (about three months later), WTI's warrants would enable it to purchase a fixed number of shares of stock at a slightly lower price than whatever the venture capitalists paid in the new round. However, if Thefacebook failed to make the deadline. WTI's warrants would enable it to pay much less—something much closer to what Peter Thiel (and WTI itself) had paid in the earlier investment round. The $300,000 Parker borrowed this way paid for most of the servers the company bought that spring as membership burgeoned.

In late March, as talks with the Post continued, Viacom entered the picture out of the blue. It expressed interest in buying the entire company for around $75 million. It wanted to combine Thefacebook with MTV.com. That was a complete surprise to everyone at Facebook. The overture was testimony to how much buzz was starting to surround Thefacebook. If Zuckerberg accepted such an offer, he would have put about $35 million in his pocket for a year's work. But that didn't matter to him. He had no interest in selling. Nonetheless, the existence of such an offer took a while to digest. At least one company adviser urged Parker to take it. He and Moskovitz would have gotten close to $10 million each.

After some back-and-forth, the Post in late March sent Thefacebook a term sheet for a very rich deal. It would invest $6 million for 10 percent ownership, after which Thefacebook would carry a value of $60 million. That's what's called in VC lingo "$54 million pre"—or before the investment: $60 million minus $6 million. Parker was thrilled. This was better than he'd hoped for. He called up adviser Conway in San Francisco. "My God! $54 million pre?" the excitable Conway shouted into the phone. "Take it! Close that sucker!" But Parker and Zuckerberg weren't in any rush. Zuckerberg's well-connected new aide Matt Cohler was urging his colleagues to keep talking to venture capitalists. Many started calling him at all hours seeking a chance to invest as soon as they heard he'd joined Thefacebook.

In any case, there was still haggling to be done with the Post. Its negotiators wanted a board seat, but Zuckerberg and Parker didn't want to give it to them unless it would be held by Graham himself. Graham thought that would be inappropriate given that his newspaper, with which he was then closely involved, might be covering Thefacebook. Zuckerberg talked it out with him by phone and they pretty much wrapped up a deal that didn't include a board seat.

The Accel Partners venture capital firm in Palo Alto was looking for a big new score. It had done a number of hugely successful investments in the previous decade. It had made its mark in the 1990s with a series of big telecommunications and software investments that paid off, like UUnet, Macromedia, RealNetworks, and Veritas. Now Internet opportunities were reemerging, but it didn't have a major consumer Net play. Some in Silicon Valley were muttering that Accel had lost its mojo.

In the aftermath of the dot-com crash of 2001–2002, Accel had reduced the size of its funds, returning money to investors unused. But by the end of 2004 it was again raising money—a new fund of $400 million. Yet some of its longtime investors were unhappy that Accel was continuing to charge more for its services than most VCs—a 2.5 percent management fee and 30 percent of any profits. Among those who decided not to continue with this latest fund was the endowment fund of Harvard University. As things turned out later, Harvard would thus end up without a stake in the creation of one of its most entrepreneurial former students.

Jim Breyer, Accel's co-managing partner, was eager to prove the firm's mettle to its latest investors. Optimism was returning to Silicon Valley, and now was the time to start taking risks again. With thick black hair, a back slapping sense of good cheer, and piercing blue eyes with which he often wryly yields a sort of knowing wink, Breyer is by nature upbeat. He likes to laugh and to share a confidence. He is a lover of music ("from Bach to Nirvana," he likes to say) and art (he collects paintings by everyone from Picasso to Gerhard Richter). And his bona fides as a master of the universe are impeccable. He is deeply embedded in the elite of American businesss—a member of the board of directors of Wal-Mart, no less. Breyer, the largest investor in Accel, knew the Internet was turning around.

Kevin Efrusy was a principal at Accel, not yet a partner—a venture capitalist in training. His mandate from Breyer: get out there and find Internet companies that could grow huge. The firm's latest interest was social networking. Accel methodically identifies the central trends at any given time, and creates a short list, which it calls its "prepared mind initiatives." One of the three at this point was "social and new media applications." But the problems with Friendster and other networks made the entire business seems risky. "Social networks sort of had this dirty name," says Efrusy, referring to a reputation for sexual material and rambunctious members. "But we'd talked about whether it was possible to keep one clean and relevant to certain demographics." He's a big, balding guy with the friendly but intense and slightly aggressive manner not uncommon among venture capitalists.

Breyer had gotten Accel close to an investment in a company called Tickle, which had been shifting its business from offering clever quizzes like "What breed of dog are you?" to becoming a true social network. By early 2004 Tickle had become the second-largest social network after Friendster, with two million members actively connected to others and exchanging messages. Accel prepared a term sheet, but Tickle's board decided to sell it to job-hunt website Monster.com instead in May 2004. An Accel partner named Peter Fenton had tried to get Accel to invest in Flickr, a photo-sharing site with some social networking features. Again, it agreed on a term sheet for the deal. But before Accel could complete that deal, Yahoo swept in and bought Flickr. Then in December 2004 Chi-Hua Chien, a Stanford graduate student doing contract research for Accel, told Efrusy about Thefacebook.

The two did a little research and Efrusy got hold of an alumni email address from Stanford, his alma mater. When he got on Thefacebook he was impressed. "They essentially had the rope outside the nightclub," he explains. "The context was there for you. It was in your college. It was Facebook Stanford, not Facebook worldwide." But Efrusy saw a problem—what kind of business would it be to run a social network for people who had no money? Then a former business school classmate explained that college students were a precious demographic for marketers. It's there that critical lifelong buying habits are formed—your first car, bank account, credit card. "But at college there was no way to reach you," says Efrusy of what he learned as he continued his research. "College was a black hole. You stopped watching TV. You stopped reading the newspaper." Thefacebook might be a way around that.

Efrusy heard how quickly Thefacebook was growing. Through a friend who had interviewed for a job at Thefacebook he got an appointment to talk by phone with Parker. Parker canceled. Then Efrusy heard Matt Cohler had joined Thefacebook. He had met Cohler at LinkedIn. Efrusy called, asking for an introduction to Parker. Cohler replied politely that the company wasn't interested in talking to venture capitalists. Then a month or so later one of Accel's partners, Theresia Ranzetta, heard that Thefacebook was talking to other VCs about raising money. Efrusy decided he had to meet these guys. After all, they were right in the neighborhood. He emailed. No response. He called. Sean Parker wouldn't return his calls.

Efrusy isn't easily deterred. He learned Reid Hoffman was an investor in Thefacebook. So he asked Accel partner Peter Fenton, who was close to Hoffman, to call and ask Hoffman for an introduction. Hoffman also demurred. He said that Thefacebook guys thought dealing with venture capitalists would waste their time because they wouldn't really understand Thefacebook and wouldn't be willing to pay what it was worth. Of course, by that point Parker was already talking seriously to several VCs. He was just deliberately avoiding Efrusy. He'd heard that Accel had lost its mojo and didn't want to deal with them. Zuckerberg, for his part, was focused on the potential Post investment. He was only half aware of how much time Parker was by now spending courting VCs.

Efrusy asked Fenton to go back again to Hoffman. This time Hoffman relented and agreed to arrange a meeting with Parker and Cohler, who was doing much of the day-to-day managerial and organizational work for the financing. However, Hoffman insisted Fenton promise that Accel wouldn't try to come in with a lowball offer. Thefacebook was pretty far down the road with a possible strategic investor, he said.

Typically the entrepreneur goes to the office of the VC, deferentially seeking funds. Efrusy spoke again with Cohler, Hoffman's former employee, and invited him to bring his partners to Accel's offices. But even then the guys of Thefacebook put Efrusy off. "He was hounding us," Cohler recalls. Efrusy had been at Accel for less than two years. He hadn't yet done a major deal of his own. He needed to prove himself.

Finally, on Friday, April 1, 2005—April Fool's Day—Efrusy decided to just go to their office. Parker had said he'd be there. Efrusy didn't realize it, but his timing was superb. The Post talks had not quite concluded because of haggling over the board seat. Efrusy walked four blocks down Palo Alto's University Avenue to an office Thefacebook had just rented on Emerson Street, about a mile from Stanford. He brought along Arthur Patterson, the tall, gray-haired, patrician co-founder of Accel, who was curious to learn why Efrusy was so excited about this little start-up. They walked up a long stairway, newly spray-painted with graffiti art. At the top was a giant, suggestive image in Day-Glo colors of a woman riding a giant dog. In the large open loft space, the company hadn't finished moving in. There was more in-your-face multicolored graffiti art all over the walls, including a few nudes. The furniture was in various states of assembly. A couple of days earlier, Thefacebook had inaugurated the space with a blowout party for Cohler's twenty-eighth birthday. Half-filled liquor bottles were scattered everywhere.

Parker had said he'd be there, but he wasn't. And Cohler and Moskovitz, who were, were hardly ready for a serious financial meeting. They were struggling to assemble do-it-yourself furniture they'd bought at Ikea. Moskovitz had hit his head on a piece of furniture and his forehead was bleeding. Cohler, usually well put together, had caught his jeans on a nail. His left pant leg was hanging open and his boxer shorts were sticking out. "Hey Kevin," Cohler greeted Efrusy.

The chaos didn't deter Efrusy. He was a man on a mission. "Our meeting was just with Matt initially," he recalls. "Sean and Mark were unavailable or sick or something. So Matt took us through the business. He was very articulate about the statistics and the repeat usage rate. I already kind of knew it, but Arthur was pretty excited. Then Sean and Mark showed up—not sick, eating burritos.

"I knew they thought we'd take too long asking questions. I said 'Hey, here's the deal. I get how valuable this could be. Come to our partnership meeting on Monday, and I promise I'll either give you a term sheet by the end of the day Monday, or you'll never hear from me again. I will not drag this process out. We can move quickly.'"

Before they left, Parker did get excited about one thing. He proudly led Efrusy and Patterson into the girls' bathroom. There he pointed to another mural, this one done by his girlfriend. It showed one naked woman embracing the legs of another. Up in a tree a French bulldog puppy looked down on them. Efrusy was nonplussed. "Sean, won't this make women uncomfortable? Aren't you worried about harassment or something?" he asked. "Look," Parker replied. "I'm not going to worry about that." Then Efrusy convinced Parker to meet him for a beer the following evening, a Saturday.

As Efrusy and Patterson were walking back to their office, Patterson slapped him on the back and said, "That was great fun. Really interesting. We have to do this." Patterson was known inside the firm for his tough-minded skepticism. This wasn't like him at all.

Over the weekend Efrusy's research went into high gear. Midday Saturday he and his wife went to Stanford and hung around Tresidder Memorial Union, the campus student center. Efrusy collared students to ask them what they knew about Thefacebook. Did they use it? How ubiquitous was it, really? The answers were what he'd hoped for. "Have I heard of it? I can't get off of it." "I don't study. I'm addicted." "Everyone's on it. You stay in touch with friends at other schools. Then all the professors get on there. It's become sort of the hub of my life."

Efrusy called the younger sister of Accel's CFO, a sophomore at Duquesne University in Pittsburgh. "She's like 'Oh yeah, Thefacebook. It came here on October twenty-third.' I was like 'You know the exact date?' She's like 'Of course. We were on the waiting list for months. We were number seven on the waiting list.' I'd never heard of anything like this. She remembers the date it was turned on. There was this rabid, pent-up demand. I talked to my wife and said, 'I've got to invest in this company.'"

That night he met Parker and his girlfriend as well as Cohler at the Dutch Goose, a grungy Stanford student dive. The talk quickly turned to money. "Kevin," Parker began, "we think this is a really valuable company. You're not going to want to pay the valuation." Efrusy begged just to be given a chance. That was exactly the reaction Parker had hoped for. Efrusy again urged Parker to come Monday morning, and to bring Zuckerberg. He was not at all sure they would show up.

But on Monday at 10 A.M., Zuckerberg, Parker, and Cohler appeared. Zuckerberg wore a T-shirt, shorts, and his Adidas flip-flops. Parker and Cohler opted for the T-shirt-under-sports-jacket look. They didn't bother showing the slides they'd displayed at other VC firms. Parker did most of the talking. It was a masterfully self-conscious display of reticence, aiming to hook Accel hard. Zuckerberg himself said very little. Then they left.

"So what do you guys think?" Efrusy asked his partners. One of the senior ones piped up. "It sounds like you've got a lot of convincing to do," he said. "Okay, but let's put that aside for a second. Do you like the business?" Efrusy asked. The room was unanimous. They did. There wasn't any debate. Patterson was enthusiastic. So was Jim Breyer. But while, as usual, Parker had presented Thefacebook's pitch, Breyer had made a critical discovery while watching the boys demonstrate Thefacebook's site. "First page of the website," Breyer wrote in a note to himself, "this is a Mark Zuckerberg company. Mark Zuckerberg is the guy." Until that moment, it had not been clear to Accel—or to any of the company's prospective investors aside from Don Graham—that Zuckerberg was the decisionmaker upon whose opinion a deal would rise or fall. Efrusy had barely met Zuckerberg. During the presentation Breyer had asked Zuckerberg to talk a bit about his background and his vision for the company, and Zuckerberg talked for only about two minutes. Since the founder's presentation is usually the most compelling part of any pitch to VCs, this reticence was noteworthy and confusing.

The Accel partners' meeting quickly turned to a strategy—how could Efrusy get Thefacebook to take Accel's money? Parker had told them that Thefacebook just about had a deal with the Post, and he roughly outlined the Post's deal terms. They decided to get a deal proposal over there fast. Efrusy and Breyer had Accel's lawyer draw terms up for an investment valuing the company at the same price as the Post had but putting in slightly more money. Efrusy got it to Thefacebook that evening. Late that night Efrusy got an email from Cohler saying thank you, we're sticking with the Post. But separately, and unbeknownst to Parker and Cohler, Breyer that evening began an email dialogue directly with Zuckerberg. The VC suggested they try to get together again the following day, Tuesday.

In reality, Parker was glad Accel was so interested. It enabled him to test the waters with the other venture firms that were still in the game. Maybe Thefacebook shouldn't do the Post deal after all. The next morning he spoke with Tim Draper of Draper Fisher Jurvetson, who said he was willing to match Accel. Parker told Efrusy that when he called a few minutes later. Efrusy suggested he might go higher and threw out a few numbers. Parker is a good negotiator, shameless. And he was excited. "No way!" shouted Parker, he recalls. "There's no way we're considering that. We want one hundred pre!" Then he hung up on Efrusy. The boys, all listening in by speakerphone, giggled.

Back at Accel, Efrusy was huddling with Jim Breyer, who wanted this deal as badly as Efrusy did. Breyer had come to see Thefacebook as something unique—a company with a kind of potential that he'd rarely felt before. They really wanted this deal to happen, and they were willing to pay to get it. But if Thefacebook wanted to go with the Post, maybe there was a way to get in on it. Breyer knew the Post's Don Graham—they'd served on a company's board together. He was sitting at lunch at his favorite restaurant, the Village Pub in Woodside, talking to one of Accel's largest investors, when his assistant finally succeeded in connecting with Graham. Breyer excused himself and stepped outside.

"Don, I understand you are talking to Thefacebook about an investment. They've also come and presented to us. We'd love to figure out a way to work with you, and split the investment 50–50," said Breyer.

"I don't think I'm authorized to do that, Jim," responded Graham. "These are the terms Mark asked for. I think we already have a deal."

"I know you have an offer on the table, but I don't think you have a deal yet," replied Breyer. "We would be happy to invest together with you if you chose to do that."

As he was leaving after lunch, Breyer made a 7 P.M. reservation at the same restaurant for dinner that night. After he got back to the office, he and his partners huddled and decided to take Accel's bid considerably higher. Tuesday afternoon, Efrusy and Accel partners Theresia Ranzetta and Ping Li walked back down University Avenue and barged uninvited into Thefacebook's office, where everybody was in the middle of a meeting. Efrusy just walked in and interrupted. "He slapped the term sheet down on the table. It showed an offer for $70 million pre, with a $10 million investment. That would make the postinvestment valuation for Thefacebook $80 million. "You've got to do this," Efrusy implored. "We get this. We have full conviction about it. We will move heaven and earth to make this a successful company." A slightly stunned Parker replied, "Okay, this is worth considering." Before he left, Efrusy noticed that the office's murals had seen some slight editing. On all the most strategic spots, someone had laid the tiniest pieces of masking tape.

After Efrusy left, the young entrepreneurs looked at one another in jubilation. Eighty million? Amazing! "But what about the Post?" Zuckerberg asked. Nobody had a good answer, but they were being offered a deal that valued Thefacebook at $80 million!

But it wasn't a clear-cut decision. On one hand Graham was a believer in the company and would let Zuckerberg and Parker do what they wanted. Though nothing had been signed, Zuckerberg had come to an oral agreement with Graham for a deal at the lower valuation. If Accel invested it would be very intimately involved, which might mean less freedom. Its offices were only three blocks away. But it also might mean more Valley wisdom and connections. Parker had no attachment to doing a deal with the Post, and Matt Cohler, the most experienced of them all, felt strongly they should raise as much money as possible. The tide was shifting toward Accel. The VC firm's term sheet explicitly provided for the possibility that the Post might invest alongside them, but there was little interest at Thefacebook in that, partly because it would have involved selling too much of the company.

With a few calls Parker determined that neither Tim Draper nor any of the other remaining VCs were willing to follow Accel to this level. It was down to Accel and the Post.

That night, Jim Breyer hosted a dinner for Thefacebook's leaders at the elegant and expensive Village Pub near Breyer's home in tony Woodside, north of Palo Alto. At the table were Zuckerberg, Parker, Cohler, and Efrusy. The Pub is known for its wine list, and Breyer, a wine connoisseur, ordered a $400 bottle of Quilceda Creek Cabernet. Zuckerberg, still only twenty and below drinking age, ordered a Sprite. The point of the dinner was in part for Efrusy and Breyer to get better acquainted with Zuckerberg, who had been mostly quiet in their meetings up until then.

Breyer by now had made known his own Harvard connection—he'd gotten his MBA there. He was even on the board of trustees of the Harvard Business School. Breyer was doing everything he could to loosen Zuckerberg up. There was some serious talk about strategy, and Breyer and Efrusy reiterated in the strongest terms how much they wanted to invest and work with Zuckerberg and his team. Breyer was starting to admire the young CEO's clearheaded way of thinking about strategy and his absolute devotion to the quality and usefulness of Thefacebook's product. Yet it was clear Zuckerberg remained uncomfortable about something. Then he started to tune out.

People often think Mark Zuckerberg isn't listening to them. He has a way of saying nothing and appearing uninterested. He does not offer the body language or nods or other conventional conversational signals that tell someone he's listening. However, that usually doesn't mean he isn't listening. He's just unemotive and quietly pensive. On the other hand, there are times when he really doesn't listen. It happens when he's either bored or very uncomfortable. On those occasions, he will repeatedly, at fairly random times in the conversation, simply mutter,

"Yeah." This distinction is only apparent to people who know him well. In the middle of the dinner at the Village Pub, Zuckerberg went into nonlistening mode. Matt Cohler noticed.

Zuckerberg went to the bathroom and didn't return for a surprisingly long time. Cohler got up to see if everything was okay. There, on the floor of the men's room, sitting cross-legged with his head down, was Zuckerberg. And he was crying. "Through his tears he was saying, 'This is wrong. I can't do this. I gave my word!'" recollects Cohler. "He was just crying his eyes out, bawling. So I said, 'Why don't you just call Don up and ask him what he thinks?'" Zuckerberg took a while to compose himself and returned to the table.

The following morning, he did call Graham. "Don, I haven't talked to you since we agreed on terms, and since then I've had a much higher offer from a venture capital firm out here. And I feel I have a moral dilemma," Zuckerberg began.

Graham had already talked to Breyer, so he was disappointed but not surprised. But he was also impressed. "I just thought to myself, 'Wow, for twenty years old that is impressive—he's not calling to tell me he's taking the other guy's money. He's calling me to talk it out.'" Graham knew that even his first offer was very high for a company so tiny and so young. He felt he had no context in which to go higher. And he assumed that no matter what he said, Accel would go even higher.

"Mark, does the money matter to you?" Graham asked. Zuckerberg said that it did. It could, he went on, be the one thing that could prevent Thefacebook from going into the red or having to borrow money.

"You know that taking their money will be different from taking our money, don't you?" Graham replied. "They will have an end in mind for you, and they will try to move you toward that result. And while we don't have the network they do, and we don't have the sophistication they do, we're not going to try to tell you how to run the company." Graham says now that "if it never went to an IPO I would have been happy. But Mark said he had thought through the disadvantages of dealing with a VC, and it was clear he preferred to do it."

"Mark, I'll release you from your moral dilemma," said Graham after a twenty-minute conversation. "Go ahead and take their money and develop the company, and all the best." For Zuckerberg it was a huge relief. And it further increased his respect and admiration for Graham.

Zuckerberg had already emailed Breyer saying he would like to meet with him one-on-one at Accel's office.

Later that morning—Wednesday, April 6—Zuckerberg walked alone over to Accel on University Avenue and sat down in Jim Breyer's small conference room. He'd come to like the affable Breyer by now, even if he hadn't been able to share his expensive wine. But Parker had been schooling Zuckerberg on the details of such investments. Turning tough, Zuckerberg told Breyer he wanted some improvements. He said if Accel raised its "pre-money" valuation to at least $75 million and if Breyer agreed to join Thefacebook's board, then he was ready to sign a deal. Efrusy was a good guy and everything, but he was junior and inexperienced. "It hurt my feelings," says Efrusy, "but I understood." Zuckerberg said firmly that if Breyer would not join the board Theface-book would simply conclude its deal with the Post instead. For Zucker-berg, the ability to place a seasoned veteran Silicon Valley investor like Breyer by his side was the determining factor.

Breyer's priority was gaining more ownership. Accel tries to own at least 15 percent of companies it invests in. But Zuckerberg and Parker didn't feel the company needed any more money. This had come up at dinner the preceding evening, and they came up with a partial solu-tion. Accel could invest $2.7 million more, and Zuckerberg, Parker, and Moskovitz would each take a special bonus of $1 million.

In venture capital deals like these, the investor usually forces the ex-isting holders to dilute their ownership prior to the investment by adding a "pool" of shares that will remain unallocated, on the assumption that future employees will get some of their pay in stock options. The way it's calculated is complicated, but it has the effect of giving the VC more of the company and the entrepreneurs less. VCs typically insist that the existing shareholders of a company accept a pool of about 20 percent.

But Parker had prepared Zuckerberg for this gambit, and it was clear at dinner the night before how badly Breyer wanted to invest. So Zuck-erberg refused to accept a 20 percent dilution. The two agreed on a 10 percent option pool instead. In addition, Zuckerberg would only accept half of that being applied to existing shareholders' ownership. So some of the dilution applied to Accel's money as well. "Mark negotiated really hard," concedes Breyer.

They finally agreed on a deal that would value Thefacebook at slightly less than $98 million post-investment. Accel would invest about $12.7 million—a stunning sum for such a small company. It would own about 15 percent of the company. "I knew the price was just way too high," Breyer says now, "but sometimes that's what it takes to do the deal." Breyer agreed to go on the board but asked if he could invest $1 million of his own money. The twenty-year-old and the VC shook hands. Zuckerberg left his office, and Breyer was elated.

Before they shook hands, Zuckerberg explained that there might still be a small additional investment coming, either from Graham and the Post or from Edgar Bronfman, the scion of the Seagram liquor fortune and the CEO of Warner Music. Bronfman had met Parker and Zuckerberg when they'd visited Warner Music in Los Angeles the previous fall, and despite the earlier animosity dating from Napster days, he and Parker had become friendly. Bronfman decided not to invest, though. (Had he in fact done so, his $300,000 in stock would be worth at least $20 million today.) The potential Viacom purchase offer was also still vaguely in the air, though Parker and Zuckerberg made no move to pursue it.

It took a few weeks before it was all finalized. Parker tweaked several key points. He further solidified the corporate structure that guaranteed that Zuckerberg controlled one unfilled board seat in addition to his own, and Parker occupied another. That meant that even with Breyer joining Thiel on the board, the two employees would command three out of the five board seats—a majority of votes. A complicated arrangement tied their stock ownership to their board seats. There would thus be little likelihood that Zuckerberg could lose control of his company. That Accel agreed to this is further testimony to how badly Breyer wanted this deal. It also attests to his faith in Zuckerberg. By now Breyer had become a believer in the young CEO, whom he says he already thought of as "a product genius." Some of his colleagues thought Zuckerberg might ask his father to take the empty seat. He had often instant-messaged him for advice during the funding process. But for now, the seat stayed empty.

Several aspects of the financing of Thefacebook were unusual. First, the sheer size of the valuation was unprecedented for an Internet start-up. Even Google's first big investment valued the company at less than $75 million. The bonus payments to the three young men at Thefacebook were kept quiet, partly because it's generally considered best in such situations if the company gets all the funds for its own purposes. In fact, so rare is such a bonus for company founders that Silicon Valley veterans cannot recall another one. As for the bet with lender WTI, "I won," chortles Parker. WTI's warrants ended up costing ten times as much as they would have if WTI's Werdegar had won the bet. But he remained a fan of Parker and the company. In July Werdegar extended yet another loan to Thefacebook, this time for $3 million, again exclusively to cover the cost of computers and other hard assets.

When Eduardo Saverin, Zuckerberg's erstwhile partner in founding Thefacebook, heard the terms of the Accel deal, he hit the roof. His share in the company, which in the summer had been 34.4 percent, had now been diluted by the additional investments and restructuring to below 10 percent. He claimed he hadn't realized this was going to happen, threatened to sue, and so forth. But since the reorganization he didn't have much leverage. Ezra Callahan had by now learned how to do all Saverin's advertising work. An outraged Saverin stopped doing any work for Thefacebook (though he kept his stock). Zuckerberg turned off his email, and Y2M was instructed to have no more to do with him.

Zuckerberg, Cohler, and Moskovitz were in awe at Parker's prowess as a negotiator. It had been a textbook case of fund-raising success. Looking back, Cohler says, "Parker was absolutely the lead on those negotiations. People now don't realize how important he was to the company. He did an outstanding job." Zuckerberg has told friends he has never seen a more amazing sales job than the one Parker did with Accel.

The day of the biggest deal in Mark Zuckerberg's young life ended in frightening anticlimax. He had signed the papers to close the Accel investment. He was now a millionaire. But late that night his impulse to keep celebrations to a minimum was almost absurdly reinforced.

Zuckerberg's girlfriend at the time was a student at Berkeley. In the wee hours he headed there to see her. On the way he stopped in East Palo Alto to get gasoline for "the Warthog," his shiny new black Infiniti. This neighborhood was much poorer than the rest of Palo Alto. The gas station was desolate. As he filled his tank a young man approached him, holding a gun. But he was so drunk or drugged he could barely stand. He had trouble speaking clearly enough to demand money. A terrified Zuckerberg took a calculated risk. He just got into his car and drove away. Nothing happened. "I feel like I'm pretty lucky," he says. Though he referred to his escape from the gunman, it's a good general observation about creating Thefacebook, and his new funding.

Finally Thefacebook had plenty of money. Now it could build a real staff. No longer would the servers be strung together with baling wire. The real growth was about to begin.

6

Becoming a Company   "Being CEO in a company is a lot different than being college roommates with someone."

Suddenly there seemed no limit to what Thefacebook could achieve. Money had been removed as an obstacle. The service continued to grow rapidly among students. Any lingering doubts Zuckerberg had about Thefacebook had been vanquished. Now was the time to make it into a real company! But wait—how do you make a company?

Mark Zuckerberg and Dustin Moskovitz were still only twenty-one years old. For all their vision, creativity, and commitment, they retained the mind-set of college kids. They knew next to nothing about how to organize a business. Sean Parker, twenty-five, had been in several start-ups but detested their restrictions and was an instinctive rebel. His willful disregard for business conventions was as thorough as Zuckerberg's ignorance of them. That left Yale graduate Matt Cohler, twenty-eight, by default the old man and official level head of Thefacebook's inner circle. He had been a McKinsey consultant and a jack-of-all-trades at LinkedIn for veteran entrepreneur Reid Hoffman, so he had a pretty good idea of what start-ups were supposed to do. But this was not an ordinary company. It did not face ordinary challenges.

Facebook's first priority was hiring more people. Now there was money for it. But people weren't sure they wanted to work at Thefacebook. Most in Silicon Valley in early 2005 still saw social networks as faddish, despite the success of MySpace. It was unclear if they could ever be businesses. What seemed hot for Internet companies around this time was blogging and podcasting. And since Thefacebook was a closed network, any adult that Thefacebook wanted to hire couldn't easily get on the site to check it out.

Compounding these problems, the still-nascent company already had a reputation for rambunctiousness. Cohler, who quickly turned his attention primarily to hiring, tried to convince a well-known recruiter for start-ups named Robin Reed to help the company find a vice president of engineering. Reed, middle-aged with short blond hair framing a round face and a New Age propensity to wear wooden beads wrapped around her wrist, wasn't interested. "I had heard wild stories about them. It was too woolly for me," she says. "Sean Parker was quite notorious at the time." Parker's reputation for partying and his forced departure from Plaxo had stereotyped him in the Valley as a bad boy. Reed talked to friends who had already tried and failed to help Thefacebook with recruiting. "It's *Lord of the Flies* over there," one told her.

While that—whatever it meant—was an exaggeration, a bunch of college kids was definitely in control. Zuckerberg had to be careful which business card he handed out at business meetings. He had two sets. One simply read "CEO." The other: "I'm CEO . . . bitch!" Not only were college administrations all over the country up in arms over the planned national beer pong competition, but closer to home, Tricia Black refused to join the company if it wasn't canceled. Black, the saleswoman from the Y2M advertising firm, had been begging Eduardo Saverin to hire her since mid-2004. She finally got her offer to join Thefacebook and set up an in-house advertising department. As for the office art, Parker's girlfriend's painting of the naked women and bulldog in the ladies' room got painted over soon after Accel invested its $12.7 million.

Recruiting tactics weren't very professional. The main method initially was a wooden figure of an Italian chef out on the sidewalk. He held a chalkboard that displayed, rather than varieties of pizza, a list of job openings like "VP of Engineering."

Cohler's first hire was Steve Chen, a former PayPal programmer. But after only a few weeks Chen decided to leave to start a new company with two PayPal friends. It was to be a video start-up, and Cohler tried to dissuade him. "You're making a huge mistake," Cohler said. "You're going to regret this for the rest of your life. Thefacebook is going to be huge! And there's already a hundred video sites!" Chen went ahead and left to start a company called YouTube.

It quickly became apparent to Zuckerberg that Google, at the top of the Silicon Valley food chain, was Thefacebook's primary competition for talent. After all, that was where just about every great software engineer aspired to work. Those were the people Thefacebook should be hiring. Simply learning that someone was interviewing at Google made Zuckerberg want him more.

Cohler visited his younger brother at Princeton and heard about a Google recruiting meeting there. He printed out a bunch of flyers about Thefacebook and stood by the door handing them out. Not long afterward Zuckerberg set up a card table in a Stanford computer science department building with a sign reading WHY WORK AT GOOGLE? COME TO THEFACEBOOK. Even Adam D'Angelo had to be convinced not to take a summer internship at Google he'd been offered. Parker persuaded him to rejoin Thefacebook instead.

After Accel made its big investment, Kevin Efrusy, who had spearheaded the deal, began visiting regularly to advise Zuckerberg. He proposed bringing on a part-time consultant named Jeff Rothschild, who had co-founded the big business-software company Veritas. Rothschild had both a deep knowledge of data centers and the maturity of a fifty-year-old. Zuckerberg realized that Rothschild could help Thefacebook prevent a Friendster-like breakdown. Efrusy suggested offering him some stock in exchange for a day a week. "Would he work full-time?" Zuckerberg asked. "No, never. He's retired," Efrusy replied. The next time the two got together, Zuckerberg proudly announced, "I got him full-time." Telling him something couldn't be done was like holding up a red flag to a bull. "I thought these guys had created a dating site," Rothschild says, recalling how fascinated with Thefacebook he quickly became. "But once I understood Mark's vision, I realized this wasn't like MySpace. It had nothing to do with meeting people. It was the most efficient way to stay in touch with your friends." Getting veteran Rothschild on board helped legitimize the company.

Rothschild's acceptance of Thefacebook helped Cohler persuade Robin Reed to help with the recruiting. She finally agreed to come see Zuckerberg. When she walked up the graffittied stairs to Thefacebook's Palo Alto offices at the appointed hour of 11 A.M., she found the door open and the place deserted. After a while she left. Cohler spotted her on the street and brought her back. It turned out Zuckerberg had been there all along—on *the roof*. If you stood on a table in a room they called the dorm room (it also had an Xbox and a futon), then shimmied out a window, you arrived at a big flat area on the roof covered with gravel. Beach chairs were set up. It was a favorite spot on sunny days and a place to get a little privacy for phone calls or meetings. Reed climbed out. Zuckerberg pleaded with her to undertake the engineering job search. She found the whole scene quite charming and finally agreed.

But Thefacebook had its own unique criteria for hiring. For one thing, there was a strong bias toward youth. Leaving school was considered a virtue among this crew of dropouts, iconoclasts, and autodidacts. "Why would you study it when you could be doing it?" Zuckerberg would ask graduate students he was trying to recruit. He even began to guarantee that the company would pay someone's tuition if they quit school to come to Thefacebook and later decided to go back. Cohler advertised for summer interns, then sometimes told promising applicants when they came for an interview that Thefacebook was only hiring full-timers. That forced people to consider dropping out. That's how he got Scott Marlette, a top early hire, to quit Stanford graduate studies in electrical engineering.

Adam D'Angelo, tall and soft-spoken with mussed-up hair and the concave posture of an introvert, remained the house egghead and programmer par excellence. He had long since stopped focusing on Wirehog. Now he was Thefacebook's top engineer. No matter who else they'd spoken with first, Zuckerberg always wanted candidates for important tech jobs to interview with D'Angelo. If he thought they were smart, they'd get hired.

When somebody did get hired, their first responsibility was to go buy their own laptop. There wasn't enough furniture, either. Scott Marlette sat on the floor his entire first week. Only two small tables stood in the middle of the main room, already tightly packed with other people's stuff. Later Marlette visited Ikea to buy his own desk and chair.

The site went from 3 million users in June to 5 million by October 2005. This was unbelievable growth, but even as they celebrated it, the growing Facebook staff had to work hard to keep it from destroying them. Their technology had to grow as quickly as their membership. Fighting the Friendster curse was a constant obsession. Adam D'Angelo became consumed with daily crises. He remembers what it was like: "This database is getting overloaded. We need to fix that. You can't send email. Fix that. This week we came this close to the edge. Next week it's going to hit the edge and the site's not going to work. We have to raise capacity." There were frequent drives down to the data center in Santa Clara to plug in more servers. By year-end Facebook had spent $4.4 million on servers and networking equipment in its data centers.

In the hectic efforts to keep everything operating despite all the rapid growth, many of the young engineers made serious errors. Some risked bringing down the entire site, since its underlying software code consisted of one very long file of instructions, violating elementary design protocols for such a project. (Marlette and D'Angelo later broke the code up into a more conventional segmented structure.) At one point, source code—the company's elementary intellectual property—was streaming onto student profiles. One engineer accidentally introduced a bug that briefly enabled any user to log into any account. On another occasion a summer intern made a coding error that meant that no matter which ad on the site you clicked, you were directed to only one advertiser—Allposters.com.

Dustin Moskovitz was responsible for keeping things running smoothly day to day, so anytime one of these disasters hit it was his job to remedy it, even if doing so took all night. When confronted with a particularly stupid error he sometimes lost his cool, banging angrily on his desk and throwing things. But he always got things fixed. He commanded universal respect for his dedication and work ethic. "Dustin was always the rock," says Ruchi Sanghvi, a beatific Carnegie-Mellon computer engineering graduate with a round face and long black hair. She was the first female engineer hired by Thefacebook and for years the only one in the company's inner core.

Rothschild, trying to figure out who was doing what, discovered that all of Thefacebook's customer support was being done by a student at Berkeley working part-time from home. The student had a backlog of 75,000 customer support requests. Rothschild advertised on Thefacebook for a customer support representative and hired a recent Stanford grad named Paul Janzer. The two quickly concluded they needed a larger staff. They rustled up six more applicants. Rothschild then held a group job interview and hired them all. Even still, the queue of unanswered requests grew to 150,000 before it started dropping. People had questions about everything from how to change their profile picture to whether they could change their name once they got married.

Efrusy tried to play the role of company conscience. In return Zuckerberg gave him a business card with the title "Chief Worry Officer." But he had cause to worry. New features weren't tested before they were inaugurated. He found it nerve-racking to sit casually conversing with Zuckerberg while he typed away on his laptop making live changes to the site.

Recruiter Reed was taking longer than she expected to conduct her search. For one thing, experienced engineers didn't like the idea of working for a twenty-one-year-old who'd never held a real job. And many interested candidates were daunted to discover Parker's reputation, especially once they learned that his title at Thefacebook was president. It also was unclear to Reed exactly what Zuckerberg wanted. His description of what kind of person he wanted kept changing. He made it clear that he himself would remain in charge of product development. But despite the seeming immaturity and enveloping chaos of daily life at Thefacebook, Reed noticed that things seemed to keep moving forward.

Zuckerberg asked her to come and work with the company full-time for six months or so, until it filled out its staff. She had never done that before, but she liked the idea of stock options in Thefacebook. She was becoming a believer. "I thought I'd stood at the elbows of a lot of great entrepreneurs and knew how it was done," she says. "But when I got to Thefacebook I was struck by how much I didn't know about how twenty-somethings worked. Everybody called them irresponsible. They didn't come in until late. Some worked only at night. But Mark was actually incredibly responsible. All of them were. So I decided to forget what I knew and have a beginner's mind."

Reed, a Buddhist and meditator, sat in the cafe of San Francisco's Museum of Modern Art one Saturday morning with Zuckerberg and made a deal. She would join the company for a few months and he would agree to meditate. She was starting to take a kind of motherly interest in his managerial success. She gave him special software for his computer that came with little biofeedback monitors that clip on your fingers and are connected by wires to your computer to measure whether you have calmed down. When they finished and came to an agreement, Zuckerberg said, "I think it's time for a hug."

Though the stresses on him these days were legion, Zuckerberg didn't seem freaked-out. In fact, he remained peculiarly placid. Even in these most hectic company days, he never lost his temper. (And he shortly reported back to Reed that he was actually using her meditation-assistance apparatus, to good effect.)

This outward placidity is one key to the peculiar charisma that both draws people to Zuckerberg and vexes them. Not only is he unemotional; he seldom betrays his feelings. His typical way to listen is to stare at you blankly, impassively. It is never obvious whether he hears you. He seldom gives any reaction to what someone says to him right away. If you need to know what he thinks, you may be out of luck. "He's really hard to read," says Chris Hughes, the former roommate who during this period was managing PR for the company from his Harvard dorm room. "It's difficult to have sort of basic communication with him."

Thefacebook had stopped being small enough for everyone to know what was going on. Now Zuckerberg had to focus more consciously on communicating, making sure his messages were passing down through the growing number of layers. Efrusy urged Zuckerberg to write down his thoughts about strategy and process. The next week Zuckerberg brought to their meeting a little leather-bound diary. "It looked like what Chairman Mao would carry around," says Efrusy. "He opened it up and it was page after page of tiny two-point handwritten text." Zuckerberg's handwriting is extremely precise, like that of an architect or designer. But he refused to let Efrusy read his notes. "I told him the point was to communicate to everybody else," says Efrusy. "He sort of looked at me like that was a new thought and said, 'Oh really?'"

This book was held closely by Zuckerberg, but some colleagues did get a peek at it. It revealed in detail where he was hoping to take his company. On its cover page it listed Zuckerberg's name and address, with a note: "If you find this book, return it to this address to receive a reward of $1,000." It was titled "The Book of Change," and just below that was a quote: "Be the change you want to see in the world—Gandhi." Inside, in Zuckerberg's precise and beautiful cursive script, were lengthy, detailed descriptions of features of the service he hoped to inaugurate in coming years—including what would become the News Feed, his plan to open registration to any sort of user, and turning Thefacebook into a platform for applications created by others. In some sections it became almost stream of consciousness, according to those who have read it. Even Zuckerberg occasionally notes in the margin, "This doesn't seem to be going anywhere." But for many of those who read it inside the company it seemed as weighty as Michelangelo's sketchbook.

A major new figure joined the life of the company around this time—investor and entrepreneur Marc Andreessen, who became a close adviser to Zuckerberg. Andreessen, one of Silicon Valley's most revered innovators and entrepreneurs, had come to California as a mere boy, much like Zuckerberg, after he helped invent the first Web browser at the University of Illinois. He co-founded Netscape Communications and later two more important and successful companies, while investing in scores more. Matt Cohler and board member Peter Thiel introduced Andreessen to Zuckerberg, thinking he could help the young CEO figure out how to grow Thefacebook. Zuckerberg immediately took a liking to the tough-minded Andreessen, who never showed the slightest hint of obsequiousness. He was consummately confident himself, in fact, never suffering fools very well. He didn't care what people thought of him, and Zuckerberg liked that. He was as blunt with Zuckerberg as he was with everyone else.

Prodded by Parker and Cohler, as well as Andreessen and Efrusy, Zuckerberg started trying to behave like a leader. He had been living in one of the company houses, but moved out in midsummer. Around the same time he proclaimed he would stop writing software. Zuckerberg needed to start focusing on bigger issues. There was a little ceremony to mark the day he installed his last piece of code. At a talk he gave at Stanford shortly afterward, he conceded, with a hint of disappointment, that "the dynamic of managing people and being CEO in a company is a lot different than being college roommates with someone." Some weekends Cohler, Moskovitz, and Zuckerberg could all be found reading Peter Drucker, the consultant and teacher often called the father of modern management.

Zuckerberg decided to study his newfound management idol—Don Graham. He asked if he could come visit the Post to observe how Graham worked. Even though at this point he barely knew the difference between profit and loss, he wanted to see what a CEO does. Zuckerberg flew to Washington and spent four days with this mentor. He shadowed Graham for two days at headquarters, then flew with him up to New York to watch him make a presentation to financial analysts. The Post Company's stock is divided into public shares and a separate class of shares controlled by the family with significantly enhanced voting power. It's a structure intended to reflect the unique sensitivities of a public company that runs a newspaper—making it a hybrid of a for-profit enterprise and a public trust—and it gives the Graham family effective veto power over company decisions. Because of this family control, Graham has the ability to enforce a long-term view. Zuckerberg started thinking he might someday want a structure like that for Thefacebook.

Zuckerberg had to figure out how to respond to the people-management challenges that arise in any organization. His approach sometimes was to make a joke out of things others might treat more gravely. A young woman complained to him that an employee had harassed her in the lunch line. His response was to publicly embarrass the perpetrator in front of everyone. "It has come to my attention," he announced at a company meeting, "that one of you said to a girl, 'I want to put my teeth in your ass.'" He paused. The room was silent. "So, like, what does that even mean?" Everyone laughed. Then the matter was dropped.

The corporate culture was an institutionalized, dormlike casualness, oddly fused to intense devotion and exertion. The twenty or so employees moved in packs—to the nearby Aquarius Theater, where they could get in for free because one of the engineers worked there part-time; to the McDonald's a few miles away in East Palo Alto; and to the University Cafe around the corner, which was the unofficial company meeting room. "We worked here all the time," says Ruchi Sanghvi. "We were each other's best friends. Work was never work for us. We worked through Christmas, over the weekends, and until five in the morning." She herself worked so hard that one night driving home in the wee hours to her apartment in San Francisco she twice ran into the center divider before pulling over and falling asleep on the side of the highway. After that she moved close to the office. Thefacebook offered a $600-a-month housing subsidy to those who lived nearby in Palo Alto, which encouraged the conflation of work and personal time.

Hardly anyone came in before noon. In his self-published *Inside Facebook*, Karel Baloun, a company engineer back then, writes that Zuckerberg himself set the tone: "Zuck would come into the office and, seeing every chair full, just lie down on the thin carpet on his belly, sandals flapping, and start typing into his little white Mac iBook." The place settled into a productive rhythm only in the evening. Programmers fueled by Red Bull would tap steadily into laptops while conversing via instant message. Fiftyish recruiter Reed started staying up at home until 3 or 4 A.M. to engage in the late-night IM back-and-forth. She realized that was when many of the important decisions got made.

Zuckerberg preferred instant messaging, using AOL Instant Messenger (AIM). One employee a few years older who sat about six feet from Zuckerberg in those days received an IM from the boss. "Hey," it read. It was the first time he'd gotten such a message. So, seeking to be convivial, this guy stood up from his chair, turned to Zuckerberg, and said out loud in a friendly voice "Hey!" Zuckerberg continued staring blankly at his screen. It wasn't even clear if he had heard. If you wanted to communicate, you IM'd. Zuckerberg became slightly more animated in the evening when many people had left.

Thefacebook made explicit efforts to be a cool place to work, almost to the point of caricature. Appearance mattered. When Jeff Rothschild started at Thefacebook he dressed like a typical middle-aged, nerdy Silicon Valley engineer—clunky running shoes and a shirt tucked into khaki pants or loose jeans. About a month later a friend ran into him at the airport. He was wearing designer jeans with his shirttail untucked in the hipster manner. "Jeff, what happened?" his friend asked. "They said I was making them look bad," Rothschild replied. "They weren't going to let me back into the office." The other employees started calling Rothschild "J-Ro." "Part of our company mission was to be the coolest company in Silicon Valley," says Parker. "I played up the idea that this should be a fun, rock-'n'-roll place to work." That's why he hired graffiti artist David Choe to paint the office and had his girlfriend add a little special something in the ladies' room. (Choe got a tiny bit of stock for his efforts, now worth tens of millions.) The company continued to rent several houses that employees shared, one of them walking distance from the office. Everybody partied there on weekends.

Zuckerberg headed to New York to meet with the new ad salesman Tricia Black had hired, Kevin Colleran. Colleran had previously worked in the record industry, and the photo on his profile showed him beaming at a party with his arm around the shoulder of rapper 50 Cent—who was goateed, impudent, and decked in bling. Zuckerberg arranged to meet Colleran in front of the Virgin Megastore on New York's Union Square. Colleran showed up late and was walking toward Zuckerberg when he got a phone call from his new boss. "Where are you?" Zuckerberg asked. "Zuck! I'm right in front of you!" Colleran replied. Zuckerberg looked crestfallen. He thought Thefacebook's ad salesman was the tough-looking black guy in the profile photo.

The unique social functions of Thefacebook were occasionally deployed on its behalf. On the day a new Stanford grad named Naomi Gleit started work at the company, Matt Cohler asked her to get all her sorority sisters to poke Jim Breyer on his profile page. It was a way to keep the board feeling good about the product.

Zuckerberg's dry wit and classicism showed then, too, according to author-engineer Baloun. "Around the end of May 2005," he writes, "Zuck painted the word 'Forsan' on his office wall in huge letters. . . . The word comes from Virgil's *Aeneid*: 'Forsan et haec olim meminisse iuvabit,' which can be loosely translated to read 'Perhaps, one day, even this will seem pleasant to remember.'"

More and more technology and media companies were taking note of Thefacebook's torrid growth and trying to figure out how they could get a piece of it. In the spring, MySpace founders Chris DeWolfe and

Tom Anderson had come up to Palo Alto from Los Angeles to put out feelers about possibly buying Thefacebook. Zuckerberg, Parker, and Cohler met him in a University Avenue coffee shop, but only because they thought these were interesting guys and they were curious about MySpace. Then, in July, MySpace itself was bought. Rupert Murdoch's News Corporation purchased the social network's parent company for $580 million to get MySpace and its 21 million users.

At Thefacebook there was celebration. Not only did the deal proclaim that services like theirs were important and valuable, but they were pleased to think that a big old-line media company would now be mucking around with MySpace. They presumed News Corp. would slow it down tremendously. Parker called DeWolfe and his partner Tom Anderson that day and put them on speakerphone so everyone could hear. The team in Palo Alto was expressing condolences. Since DeWolfe and Anderson didn't own much of the parent company, they weren't going to get much money from the sale of their creation.

The last thing Zuckerberg wanted to do was to sell his baby. At his Stanford talk, someone asked him what he thought might be the best way to "monetize" or make money from Thefacebook, "as an exit strategy." Zuckerberg's reply was his only curt one of the night. "I spend my time thinking how to build this and not how to exit," he replied. "I think what we're doing is more interesting than what anyone else is doing, and that this is just a cool thing to be doing. I don't spend my time thinking about that. Sorry."

Though Zuckerberg put minimal priority on advertising, a fair amount was coming in anyway. But even at this early date it was apparent that Thefacebook wasn't like a typical website when it came to advertising. That was both a good and a bad thing. For one thing, ads on Thefacebook didn't get clicked very often. Some believed that was because when users were focused on finding out about friends they were unreceptive to commercial messages. A version of the Google model, which charges advertisers only when their ads are clicked on, did not look promising here.

Colleran, the new ad salesman in New York, was working hard to find brand advertisers who would pay on the basis of CPM, or cost per thousand views. That's how television ads are priced. The goal of this kind of ad (as opposed to the pay-per-click ones that Google specializes in) is not to get clicked, but rather to be seen by lots of people. But Thefacebook was still an exotic site for college kids that few on Madison Avenue had heard about and even fewer understood.

For some months Colleran was the site's only full-time ad sales-man, and he quickly became frustrated. This big chummy guy with a blond crew-cut was a gung-ho cold-caller who could get in almost any door. He turned up plenty of advertisers willing to try Thefacebook. But many of their ideas were rejected out of hand by Zuckerberg. He vetoed anything that smacked of interference with the fluid use of the site, no matter how much revenue it might generate. Common practices like pop-up ads that displayed before you saw the content of a page were ab-solutely anathema, for example. Colleran learned to be cautious about what he even suggested.

It drove Colleran crazy that Zuckerberg wouldn't add new schools to Thefacebook more quickly. To the ad guy, more users just seemed better. Zuckerberg and Moskovitz, however, were methodical. Students from schools where Thefacebook hadn't yet launched regularly came to the site and tried to sign up. They would go on a waiting list and be alerted when it came to their school. When the number on the waiting list passed 20 percent or so of the student body, Thefacebook would turn that school on. "I always thought it was wrong," says Colleran, "but now I realize it was a major reason for our success." By keeping the gates closed and only opening at schools once there was proven demand, Zuckerberg and Moskovitz, the expansion guru, ensured that when Thefacebook did open, usage would explode.

Colleran found one company willing to pay big bucks for ads. It was his first big deal. British online gaming company Party Poker didn't buy its ads on a CPM basis, but rather on what's called CPA—cost per acquisition. Party Poker paid a flat fee of $300 for each new subscriber who signed up for its service and put at least $50 into a gambling ac-count. This proved to be hugely lucrative for Thefacebook—it was soon reaping $60,000 each month just for the 200 new members, on average, who signed up. Salespeople at Y2M, which was also still selling ads for

Thefacebook, were astonished. They had never before seen a college advertiser spend so much on the Internet. A year or so later, though, online gambling was outlawed in the United States and Thefacebook dropped Party Poker.

Those interested in banner ads on a CPM basis included some companies that targeted college students as employees—like house-painting operations and door-to-door retailers that hired students for summer work. One big client sold kitchen knives. Businesses that provided products sold by fraternities and sororities for fund-raisers also saw a good response on the site. Ads started at $5 per thousand views and advertisers had to spend a minimum of $5,000 per month.

But aside from the lucrative Party Poker deal, the main revenue was still coming from sponsored groups, especially Apple's. Since Apple paid $1 per month per member, as the Apple group grew Thefacebook made more and more. Soon it was generating hundreds of thousands of dollars per month. That was the single biggest source of revenue the company had in 2005. Other companies that sponsored groups, which required a minimum monthly payment of $25,000, included Victoria's Secret.

But there were also early signs that this new kind of social network offered uniquely powerful tools for advertisers. In 2005, Interscope Records released a single by Gwen Stefani called "Hollaback Girl." The song takes the form of a sort of cheerleading chant, and Interscope's marketers got the idea of promoting it explicitly to college cheerleaders, hoping they would adopt it for their routines at games. Where better to find college cheerleaders than at a college-only website? Dustin Moskovitz had become good at mining profile data on Thefacebook for advertisers, so it was little trouble for him to target cheerleaders.

This approach might seem obvious, but few sites on the Internet before this could offer targeting based on information that had been explicitly provided by users. Interscope could have instead hired a firm that targets users on other sites based on inferential analysis of Internet behavior. Such ad networks watch what people do using tiny pieces of software called "cookies," which are installed in consumers' Web browsers. They can know, for example, you have been to the kinds of sites a twenty-year-old girl might go to, or that you have shopped for pop music online. If you did both, they might place ads on pages you visit. Such an approach infers who you are and what you're interested in by supposedly savvy guesswork.

While such targeting has been considered acceptably accurate, it's a shotgun approach. Many such ads are seen by people who aren't the real targets. Even gender targeting is often inferred incorrectly online. One longtime Internet ad executive estimates gender targeting errors at 35 percent. If you are sharing your boyfriend's laptop, for example, this approach won't work very well. Another way an advertiser like Interscope could achieve tight demographic targeting would be to find a site just for college cheerleaders, if one existed, and run its ad there. But it wasn't likely to get large numbers that way.

On Thefacebook, by contrast, Interscope could be given a guarantee —its ad would only be seen by college girls who are either cheerleaders or who have mentioned something about cheerleading in their profiles. The company told Interscope exactly how many times it displayed the ad on pages seen by such girls. "Hollaback Girl" did become a popular cheerleading anthem at football games that fall. It's impossible to prove the ads on Thefacebook were determinative, but it's a fair bet that just about every cheerleader at the schools where Thefacebook operated saw them.

Targeting of this type is enormously promising. A media kit used by Colleran right after he started lists the following parameters an advertiser could use for targeting college students: geography, gender, course, keywords in profile, class year, major, relationship status, favorite books, movies or music, political affiliation, and university status (student, faculty, alumni, or staff). The house-painting and knife-peddling advertisers could show their ads only to male students at colleges in regions where they wanted to increase their workforce. Or they could aim more tightly—at freshman males on the football team who had gone to high school in northern Ohio.

Inside the company it was starting to sink in on these young pioneers that they had a unique database about people that could be tapped for many purposes. The combination of real validated identity information and extensive information about individuals could yield insights no Internet service previously had seen. A math whiz friend of D'Angelo and Zuckerberg from Exeter spent the summer writing algorithms to find patterns in Thefacebook's data. He was able to create lists of user favorites. Movies were the top interest of the service's 3 million users. Their five favorites were *Napoleon Dynamite*, *The Notebook*, *Old School*, *Fight Club*, and *Garden State*. Favorite book: *The Da Vinci Code*. Favorite musician: Dave Matthews. Soon the service began offering something called Pulse. It tracked which books, movies, and music were most popular on Thefacebook as a whole and on a given college campus.

For all the promise Thefacebook's unique data held for advertisers, most of the ads that were selling on the site at that point were generic banner ads. Facebook had contracted with several ad networks, which were posting ads willy-nilly. None of it was generating very much revenue. The company was steadily burning through the money it had raised from Accel. By year-end, it had $5.7 million left from the $12.7 million it had raised. Thefacebook had not yet become a real business.

These highly intellectual dropouts would spend endless hours debating what Thefacebook was really doing. After all, there had never been a website quite like this before. They took a serious, almost grave view of the significance of what they were building. Zuckerberg referred to it as a directory of people. That was, he said, what he had originally set out to build. Parker put it more imaginatively. He said Thefacebook was like a little device you carried around and pointed at people so it would tell you all about them. Cohler's analogy was that it was like your cell phone—a gateway to the people in your life. Even back then they often heard the criticism that Thefacebook was a waste of time. Zuckerberg's standard rebuttal: "Understanding people is not a waste of time." He started saying that the goal of Thefacebook was "to help people understand the world around them."

They loved to talk about how Thefacebook showed what economists call "network effects." And it did, just as have many of the great communications and software innovations of the last hundred years. A product or service is said to have a network effect when its value grows greater to all users each time one new user joins. Since every incremental user thus in effect strengthens the service, growth tends to lead to more growth, in a virtuous cycle. That was surely the case with Thefacebook, just as it was with instant messaging, AOL, the Internet itself, and even the telephone. Businesses or technologies with network effects tend to grow steadily and to have a durable market presence.

While they wanted working at Thefacebook to be seen as cool—that helped in recruiting—the product was another matter. Friendster had lived by its coolness and was now dying. Thefacebook, Zuckerberg began declaring, was "a utility." No term could sound more boring, though he was really thinking in grandiose terms. He meant it as a way of claiming Thefacebook's affinity with the telephone network and other communications infrastructure of the past. "We wanted to build a new communications medium," says Parker. "We knew we'd be successful when we were no longer cool—when we were such an integral part of peoples' lives that they took us for granted." Dustin Moskovitz adds that it was important for the company to escape the associations that came with its campus roots. "It was always very important for our brand to get away from the image of frivolity it had, especially in Silicon Valley," he says. He had not been a big advocate of the beer-pong tournament.

Sleekness and efficiency were the image they sought, rather than frivolity. Though Thefacebook's white and barren functional look stood in stark contrast to the florid excess of MySpace, its design was still awkward and inefficient, reflecting the additive way it had evolved since its dorm-room days. Aaron Sittig, the graphic designer and programmer who was Parker's close friend, had now joined Thefacebook full-time. "On my first day I came in and asked Mark, 'What do you want me to do?'" Sittig remembers. "And he's like 'You're a designer. So redesign the site.'" It came to be called the Facelift project. Sittig spent the summer working closely with Zuckerberg to disentangle software code and simplify how everything worked. The simplicity that later came to characterize the site was deliberate. "We wanted to get the site out of the way and not have a particular attitude," says Sittig. "We didn't want people to have a relationship with Facebook so much as to find and interact with each other."

Another major project in the summer of 2005 was acquiring the Internet address Facebook.com so the service could change its name. Parker, especially, was offended by the awkward inclusion of the article *the* in Thefacebook. He spent weeks negotiating with a company called AboutFace, which used the Facebook.com address to market software that companies used to create employee directories. AboutFace was willing to sell, but didn't want Thefacebook stock as payment. Parker ended up paying $200,000 in cash. He also oversaw a redesign of the logo, removing the brackets that had surrounded "thefacebook" and streamlining the typeface for the new company name: Facebook. The partly pixelated head of Al Pacino in the upper left corner of the screen remained, cleaned up a little and shrunk. The company officially became Facebook on September 20, 2005.

But despite Parker's successes, every day it became clearer to Zuckerberg and others that he was not the right guy to be helping manage the company. Zuckerberg started to think *he* should run the company. Parker himself doesn't deny he was unreliable. "I'm always gearing up for a really big push and achieving a lot and then kind of disappearing," he admits, "which is not a good trait if you want to be operationally involved in a company day to day." Parker was disappearing periodically. And employees noticed his erratic moods.

Rebranding Facebook would turn out to be Parker's last important act as company president. In the last week of August he was on a kiteboarding vacation in North Carolina, where he had rented a house right by the beach with several friends, including a young woman who was his assistant at the company. That she wasn't yet twenty-one would figure in Parker's later difficulties. One night midway through their vacation week, they threw a party and invited the kiteboarding instructors, who in turn invited a bunch of their local friends. The party got so big that people began dropping in off the beach. Then two nights later, the final night, they hosted another, smaller gathering with the instructors.

The group was drinking beer when a horde of police burst in with drug-sniffing dogs and a search warrant naming "Scott Palmer." They said they had a report that the house contained a large amount of cocaine, ecstasy, and marijuana. They searched everywhere.

Parker and his friends repeatedly insisted that the police were mistaken and that there were no drugs. But finally, after about an hour, a policeman triumphantly returned brandishing a plastic bag containing white powder. Parker, who had signed the rental agreement for the house, was taken to the police station. When he got there he learned there had been reports of drug use following the party two nights earlier. After a lengthy back-and-forth over whether there was even enough evidence to book him, Parker was arrested for felony possession of cocaine. He was not formally charged with a crime. That would require an indictment. He was released immediately.

Parker flew home to California, shaken but adamantly insisting he had done nothing wrong. He told Zuckerberg, company counsel Steve Venuto, as well as executives Dustin Moskovitz and Matt Cohler. They decided it didn't call for any action by the company. Then Zuckerberg told Jim Breyer about the incident. That did not portend well for Sean Parker.

Breyer, Accel's board member at Facebook, took the arrest very seriously. He was worried not only that the company's president and board member was being accused of drug possession, but also that he had been with an underage company employee at the time. Breyer knew about the allegations about drug use and misbehavior at Plaxo, because he had talked to Mike Moritz and other investors in that company about Parker before investing in Facebook.

The fact that Parker had never developed a good relationship with Accel and Breyer made it hard for him to resolve the matter quietly. A complicated and tense negotiation ensued.

Zuckerberg was not convinced Parker had done anything wrong. No official charges had been brought, after all. (They never would be.) And Zuckerberg felt real loyalty to his friend. The CEO was deeply grateful to Parker for having done such a good job negotiating with Accel, and for ensuring he had control of the company.

But Breyer thought Parker was a liability for the company well beyond his actions in North Carolina, whatever they might have been. Though he had tremendous respect for Parker's intelligence, he saw him as bringing a volatile edge to the company's culture. Breyer was also fully aware of Parker's aversion to venture capitalists like himself.

Others in the company's leadership felt uncomfortably stuck in the middle of an intractable dispute. Even some of Parker's friends felt that, regardless of the merits of this particular accusation, he should not remain Facebook's long-term president. For them, this incident was merely the straw that broke the camel's back. So while some of these younger employees supported Parker in his specific contention that he had done nothing wrong, they weren't eager to retain the status quo. Among other things, his compatriots worried about Parker's desire to remain the public face of Facebook. It felt risky for the company to have that same person leading what was beginning to seem like a reckless personal life.

A whirlwind of accusations and arguments swept the company up into a genuine crisis. Breyer insisted Parker had to leave the company. The pressure on Zuckerberg was intense. Meanwhile, Jeff Rothschild, who had been brought in as a seasoned technologist by Accel but had by now bonded with the team of young entrepreneurs, worked hard to serve as a mediator. He spent hours talking with Parker and Zuckerberg as they sought a resolution, as did company counsel Venuto (a longtime associate whom Parker had hired).

All this took place over only a couple of days. Breyer demanded Parker step down and was talking about filing a lawsuit because as a board member he hadn't been informed earlier. Parker's friend and board member Peter Thiel was also encouraging him to quit. Parker and Zuckerberg sat in the dorm room and had an emotional conversation, which ended with Parker agreeing to step down.

But this third time he was being ejected from a company he had helped create, Parker had finally succeeded in building in some insurance for himself. Under the terms he had carefully crafted to protect himself and Zuckerberg, he had no obligation either to relinquish his board seat or to give up his stock options, even if he was no longer an executive. But Breyer insisted that he not only leave the board but also stop vesting, or acquiring final ownership, of his stock, since he had only been at the company about a year. (Vesting is generally tied to tenure—the longer you remain with the company the more stock becomes yours.) The company advanced Parker's vesting by a year, and he agreed to relinquish about half his options. (Had he retained the options he gave up, they would be worth around $500 million today.)

But Parker had the right to assign his board seat, which he was also voluntarily relinquishing, to someone else. He had reservations about giving it to Zuckerberg, because with the control of a third seat Zuckerberg would have unchallengeable authority over the company's destiny. However, Parker worried that any other choice would risk allowing the company to fall under the control of outside investors. His assumption that, if investors had the power, they would eventually seek to oust Zuckerberg gave him, he felt, no choice.

Parker and Zuckerberg agreed the seat should revert to the CEO, giving Zuckerberg control of two seats on the five-person board in addition to the one he occupied himself. For the time being these two seats remained unoccupied. But in the event of any serious disagreement with Breyer and Thiel, Zuckerberg had the ability immediately to appoint two new directors on the condition they vote as he instructed. "That solidified Mark's position as the sort of hereditary king of Facebook," says Parker. "I refer to Facebook as a family business. Mark and his heirs will control Facebook in perpetuity." Zuckerberg continues to this day periodically to consult his former colleague.

7

Fall 2005 "He was formulating a broader and broader theory about what Facebook really was."

As the school year resumed in the fall of 2005, the company now named Facebook had effectively blanketed the college market—85 percent of American college students were users and a full 60 percent returned to it daily. Now Zuckerberg wanted to broaden membership into new demographics. But many in the company wondered whether it made sense. "The debate was 'What's next?'" says board member Jim Breyer. "Do we go international? Do we go young adult and keep the people who are graduating? But we knew that if we were going to win big, we had to start getting the hearts and minds of high schoolers."

Zuckerberg and his co-founder Moskovitz, for their part, saw Facebook on a slow march toward ubiquity. To them, high school was just an obvious next step. This could be a huge leap in Facebook's audience. And it was important to counter MySpace, which was making rapid inroads in high schools. Once you knew how Zuckerberg felt, you knew how the board was going to vote.

So Facebook had that summer started planning to include high school students. Investor Breyer and Matt Cohler—the older people— both argued that the Facebook brand was irrevocably associated with college and that college students didn't want high schoolers in there with them. They argued that a high school Facebook should operate separately and under a different name. Facebook High was considered promising, but "FacebookHigh.com" was owned by a speculator who wanted too much money for it.

If high school students joined Facebook, how would the service validate users? Protecting the culture of real names and genuine identity was critical. The college-issued .edu email addresses had ensured that people were who they said they were. That was the foundation that enabled Facebook to protect its users' information—you only shared stuff with people you knew. More than half of all users had so much faith in the security of their information that they included their cell-phone number in their profile.

However, only a small number of high schools, mostly private ones, gave students email addresses. New general counsel Chris Kelly, who had recently been hired, briefly launched a campaign to convince high schools to issue email addresses to students as an online safety measure. Then Facebook considered instituting its own national high school email service. Finally it came up with a compromise. Part of what authenticated you on Facebook was the people who, in effect, vouched for you by being your online friends. So college freshmen and sophomores were encouraged to invite their friends who were still in high school. Then those users could invite their own friends. It meant a slower start for the high school version of Facebook. The service created separate "networks," or membership groups, for every one of the country's 37,000 public and private secondary schools.

Initially, the high school site operated as a separate "Facebook." Though high school users also logged in at Facebook.com, they couldn't see college users' profiles. Membership grew painfully slowly at first, but by late October thousands of high school students were joining the service each day. (Overall at that point, about 20,000 new users were joining daily.)

Facebook was no longer just a college phenomenon. Zuckerberg, with the strong support of Moskovitz, soon insisted that the two services should be merged. By February 2006 they were ready to abandon that distinction, so users could freely establish friendships or send messages with anyone regardless of age or grade (the minimum age was set at thirteen). Cohler and Breyer and many of the older employees remained extremely worried that Facebook's appeal to college kids would plummet when they saw high schoolers in there with them.

So it was a very dramatic day for them when they merged the two systems. But it turned out college kids—the ones who noticed— were generally pleased to be able to communicate with a larger universe of potential friends. There was some griping as there always was when Facebook expanded beyond what was seen as a formerly exclusive cohort. One new group was called "You're Still in High School and You're Friending Me? That's Awkward . . . Now Go Away." But the data told Zuckerberg and his crew what they wanted to know. It showed that lots of communication was developing between high school and college kids and that overall activity was going up as a result of the change. By April 2006, Facebook had over a million high school users.

Facebook had outgrown its cramped warren of rooms above the China Delight restaurant on Emerson Street in Palo Alto. The company decamped for larger quarters one block away on University Avenue, not far from Stanford and across the street from Google's original headquarters. Facebook relocated to a modern glass office building, indicative of a new gravitas for the company. Moving, however, involved improvisation of the classic Facebook variety. Everybody carried their stuff themselves. A short procession ensued as a row of T-shirted, unkempt young engineers pushed their desk chairs, each one loaded with an extra-large monitor, along the sidewalk for the one-block trip.

When Facebook reached 5 million users in October 2005, it held another party at board member Peter Thiel's San Francisco club Frisson to celebrate—only ten months after the one-million-user party there. Every day brought more evidence that users were infatuated with the service. At the beginning of the school year, Facebook had nearly doubled the number of colleges where it operated—to over 1,800. At almost every one, its penetration among students quickly surpassed 50 percent. More than half of users were signing in at least once a day—an extraordinary statistic for any Internet business. And in the office, the staff was being bombarded with emailed pictures of quails.

Users had noticed the quote from *Wedding Crashers* at the bottom of the search page that said, "I don't even know what a quail looks like," and they were trying to be helpful. Or else they were in on the joke. Or both. It didn't matter. They cared.

Users were viewing 230 million pages daily on Facebook, and revenue had climbed to about $1 million per month. Mostly it was coming from ad networks that were placing low-priced display ads. Sponsored groups like the ones run by Apple and Victoria's Secret were bringing in thousands, and announcements at individual schools generated some money as well. But since the company's costs each month were about $1.5 million, Facebook was burning through its capital at the rate of about $6 million per year. The money was mostly coming out of the Accel investment, and Zuckerberg wasn't very concerned. Neither was Moskovitz. Moskovitz kept working like a dog, but when he wasn't at his desk he was driving proudly around in a new BMW 6-series sedan he'd bought in September.

There was a sense among many at the company that they were participating in something historic. Cohler, who unlike most of this crew had actually received a degree, from Yale in music, saw analogies. "It was one of those moments with a unique creative zeitgeist," he says, "like jazz in New York in the 1940s or punk in the 1970s, or the first Viennese school of the late eighteenth century." The conviction that this was history in the making led people to work even harder.

The history was not being made by Facebook alone. The company was surrounded by other companies also creating a more social Internet. Just around the corner was Ning, funded by Marc Andreessen and building software that enabled anyone to create their own private little social network. Up in San Francisco, forty-five minutes to the north, Digg was inventing a new tool that allowed people to share articles and other media they found on the Web. Other social networks like Bebo and Hi5 were emerging there, too, some targeting the same users as Facebook but in any case building clever products that were resonating with users all over the world.

Moskovitz was more interested in user numbers than historical analogies. Ever vigilant about competitors, he was worried that MySpace had grown from about 6 million members in January to 24 million by now. "How are they doing it?" Moskovitz asked one day. "Fuck MySpace," Zuckerberg replied.

He had a chance to express a similar disparaging view in slightly more polite language directly to MySpace's leaders shortly thereafter. Zuckerberg and Cohler flew down to Los Angeles, where they sat at a restaurant with Ross Levinsohn, head of Fox's interactive group for Rupert Murdoch's News Corp. He oversaw MySpace. Their competitor was being solicitous again. Levinsohn was cultivating Zuckerberg because he wanted to buy Facebook to add to his digital portfolio. But Zuckerberg was, as usual, just stringing him along. In her book *Stealing MySpace*, Julia Angwin recounts how Levinsohn seemed dubious Facebook could handle its rapid growth. Zuckerberg was dismissive, both of the comment and of Levinsohn's business. "That's the difference between a Los Angeles company and a Silicon Valley company," he said. "We built this to last, and these guys [at MySpace] don't have a clue."

A few weeks after it hit 5 million users, Facebook added a new feature that would transform its service. It had succeeded up to that point by being what one employee called "brain-dead simple"—all you were able to do was fill in your own profile and scan the information others had put into theirs. But there was one way to customize and modify your profile that had become very popular. Though you were allowed only one profile photo, students were frequently changing that photo, sometimes more than once a day. They clearly wanted to be able to post more photos.

Photo hosting was exploding on the Internet. Earlier that year Yahoo had acquired Flickr, a pioneering service that allowed users to upload photos for free, and was very creative with something called "tagging." A tag was inserted by the photographer when he or she uploaded the photo, to label it based upon its content. A single photo might be tagged "landscape," "Venice," and "gondola." Users could search for photos based on their tags.

A lengthy debate ensued about the wisdom of Facebook getting into the photo-hosting and storage business. The earlier add-on Wirehog application, which was intended partly to enable users to see photos on one another's PCs, had fallen flat. During the brief period when the Wirehog application was active, few users tried it. And Zuckerberg worried that to tinker with Facebook's simplicity was risky when the service was growing so rapidly just as it was. But finally Parker and others convinced him it was worth a try to build a Facebook photos feature. "The theory behind photos," says Parker, "was that it was an application that would work better on top of Facebook than as a free-standing application."

Some of the company's best new arrivals took on the project. Aaron Sittig oversaw the user interface and design. Engineer Scott Marlette wrote the software. Managing the process was newly hired vice president of product Doug Hirsch—the fruit of Robin Reed's recruiting labors. At thirty-four, Hirsch was an online veteran who had been one of the first thirty employees of Yahoo.

After a few weeks, Sittig, Marlette, and Hirsch quickly came up with a well-designed if conventional photo-hosting service. Like many on the Internet, it allowed users to upload photos and include them in online albums, and enabled others to comment on them. But they knew it wasn't exactly right. Hirsch, who had years of experience in Internet product design, suggested they take a different approach, something uniquely Facebook. "I wish there was just one really social feature we could add to this," he said in a meeting. Sittig, a very serious young man with blond bangs whose impeccable beach-boy good looks are seldom graced by more than a fleeting and wry half-smile, considered what that might mean. "I went back and thought a bit," he recalls, "and I was thinking, 'You know, the thing I most care about in photos is, like, who's in them.'"

It was a breakthrough. They decided that Facebook photos would be tagged in just one way—with the names of the people in them. It sounds elementary but it had never been done before. You would only be able to tag people who had confirmed they were your friends. People who were tagged received a message alerting them about it, and an icon appeared next to their name on the lists of friends that appeared on each user's page.

The photos team made two other important decisions. To see the next photo, all you had to do was click anywhere on the photo you were looking at. You didn't need to hit a little "next" button. They were attempting to encourage that "Facebook trance" that kept people clicking through pages on the service. It made looking at photos simple and addictive. They also took a gamble and decided to compress photos into much smaller digital files, so that when they appeared on Facebook they were significantly lower in resolution than the originals. That meant they would upload faster, so users could select a number of photos on their PC and see them online within minutes.

Would people accept low-resolution photos? Would they use the tags? On the day in late October when the team turned the Photos application on, they nervously watched a big monitor that displayed every picture as it was uploaded. The first image was a cartoon of a cat. They looked at each other worriedly. Then in a minute or so they started seeing photos of girls—girls in groups, girls at parties, girls shooting photos of other girls. And these photos were being tagged! The girls just kept coming. For every screenful of shots of girls there were only a few photos of guys. Girls were celebrating their friendships. There was no limit to how many photos people could upload, and girls were putting up tons of them.

Ordinary photos had become, in effect, more articulate. They conveyed a casual message. When it was tagged, a photo on Facebook expressed and elaborated on your friend relationships. "Pretty quickly we learned people were sharing these photos to basically say, 'I consider these people part of my life, and I want to show everyone I'm close to them,'" says Sittig. Now there were two ways on Facebook to demonstrate how popular you were: how many friends you had, and how many times you had been tagged in photos.

Sittig, Marlette, and Hirsch had also stumbled onto a perfect new use for photographs in the age of digital photography. More and more people were starting to carry cell phones with built-in cameras, using the cameras for quick snaps of daily activities. If you always had a camera with you, you could take a picture simply to record something that happened, then put it on Facebook to tell friends about it. The tags on a photo automatically linked it to people throughout the site. This was very different from the way photos were generally used on MySpace. MySpace was a world of carefully posed glamour shots, uploaded by subjects to make them look attractive. In Facebook, photos were no longer little amateur works of art, but rather a basic form of communication.

In short order the photos feature became the most popular photo site on the Internet and the most popular feature of Facebook. A month after it launched, 85 percent of the service's users had been tagged in at least one photo. Everyone was being pulled in whether or not they wanted to be. Most users had their profile set up so that if someone tagged them in a photo they received an alert by email. Who wouldn't go look at each new picture of themselves once they got that email? After the photos feature launched people began to come back to Facebook more often, since there was more often something new to see. This thrilled Zuckerberg, whose primary measure of the service's success was how often users returned. A full 70 percent of students were now coming back every day, and 85 percent at least once a week. This is astonishing customer loyalty for any Internet service, or any business of any kind, for that matter.

Immediately the question shifted to whether Facebook could handle all the new data and traffic. It put a massive burden on the storage and servers. Within six weeks the photos application had consumed all the storage that Facebook had planned to use for the coming six months. Having data center software veteran Jeff Rothschild on hand proved fortuitous. He stayed late night after night, trying to keep the company's servers from "redlining"—exceeding their capacity and potentially crashing. People from across the company were drafted to trek to the data center and help plug in new servers. Marlette, considered by most of his colleagues a programming genius, focused on rewriting the photo software code to make it more robust and efficient. By late 2009 Facebook was hosting 30 billion photos, making it the world's largest photo site by far.

The success of photos led to an epiphany for everyone at Facebook, from Zuckerberg on down. The team had built what was otherwise a plain-vanilla photo-hosting application. But the way they integrated it with Facebook showed the magic of overlaying an ordinary online activity with a set of social relationships.

Facebook executives were seeing the Facebook Effect in action themselves for the first time. Zuckerberg was beginning to talk about what he would come to label the "social graph," meaning the web of relationships articulated inside Facebook as the result of users connecting with their friends. With Facebook photos, your friends—your social graph—provided more information, context, and a sense of companionship. But it only worked because the photos were tagged with people's names and Facebook alerted people when they were tagged. The tags determined how the photos were distributed through the service. "Watching the growth of tagging," says Cohler, "was the first 'aha' for us about how the social graph could be used as a distribution system. The mechanism of distribution was the relationships between people."

Perhaps applying the social graph to other online activities would make them more interesting and useful, too. But how could Facebook help make that happen? If photos were a new application on top of the Facebook platform, what would some other applications be? Zuckerberg found these to be enormously exciting questions, and they dovetailed with thoughts he had discussed with Adam D'Angelo since even before Thefacebook launched about how the entire Internet needed to become more "social." It was the Wirehog dream finally coming to fruition. "Watching what happened with photos," says Parker, "was a key part of what led Mark's vision to crystallize. He was formulating a broader and broader theory about what Facebook really was."

Harvard continued to figure in Facebook's story. Following the success of photos, Zuckerberg began scheming to make more dramatic changes in the service, but to implement them he would need a bunch of new top-quality programmers. He had been frustrated by the people who were applying in Silicon Valley. They just didn't fit in with Facebook's culture. They were too corporate, not iconoclastic enough, and not in his view sufficiently creative. So he combed through Facebook looking up old teaching assistants and other computer science majors who had impressed him at Harvard. He wrote out a list and gave it to Robin Reed, who started calling them. It turned out that a bunch were living in Seattle.

In January 2006, Facebook hired four former computer science teaching assistants from the Harvard classes of '03 and '04: three worked at Microsoft and one at Amazon.com. One—Charlie Cheever, from Microsoft—Zuckerberg thought of as a kindred spirit because Cheever had been brought before Harvard's Administrative Board for downloading student information into a database. Cheever let a few friends search through his program to find out who roomed with whom, or which dormitory that cute girl lived in. It was an escapade not unlike Zuckerberg's with Facemash, but a year earlier.

This influx of programming hotshots immediately brought a new rigor and focus to Facebook's engineering. Not only were they young enough to understand the ethos of openness and transparency that was at the heart of the company's values, but they had several years of experience at the best software companies under their belt. They expected nothing less than to participate in groundbreaking Internet innovation.

8

The CEO "You'd better take CEO lessons!"

As Facebook kept evolving—and growing faster with every change—the established powers of the technology and media world began paying ever closer attention. This appeared to be the kind of irresistible consumer website every executive had dreamed of owning since the Internet took off in the mid-1990s. Mark Zuckerberg suddenly had a lot of new older, well-dressed friends from Los Angeles and the East Coast.

But he didn't think like the CEO of an established technology or media company. He barely gave a thought to profit and was still ambivalent about advertising. This wasn't easy for his newfound suitors to understand. One senior executive from a tech company recalls a frustrating visit during that time with Zuckerberg, who seemed uninterested in increasing the company's revenue. "He didn't know what he didn't know," he says. "But when he opened his mouth he was very direct, very smart, and he was very focused on Facebook as a social tool, to the point of naïveté. It sounded just too altruistic at the time. So I asked him, 'Is it a social tool as a tactic to get to the next point?' And he says, 'No, all I really care about is doing this social tool.' So I thought, 'Either this guy is being very strategic and not telling me what his next thing is, or he's just got his sandbox and he's playing in it.' I couldn't figure it out."

Viacom's MTV subsidiary had identified Facebook as a natural partner back in early 2005 when strategy boss Denmark West had vainly proffered the idea of a $75 million acquisition. A few months after that overture was rejected, MTV almost succeeded in buying MySpace, only to have it snatched away by News Corp. in July. Viacom's octogenarian CEO, Sumner Redstone, was enraged that archrival Murdoch had stolen his prize. By the fall of 2005, MTV's interest in Facebook was stronger than ever. After all, West and others reasoned, there was so much overlap between the two companies' audiences that Facebook could *be* MTV's digital strategy.

West called Cohler, who told him that Zuckerberg only wanted to have CEO-level conversations. If Viacom CEO Tom Freston would participate, Zuckerberg would come for a meeting. It was quickly arranged. Cohler and Zuckerberg flew to New York to meet with Freston and MTV Networks CEO Judy McGrath. Freston, solicitously, explained that there seemed great synergy between MTV and Facebook because of how much their audiences overlapped. He said he'd love to find a way to work together. He suggested, for example, that Viacom could help Facebook develop content for its growing audience. "We think of ourselves as a utility," Zuckerberg replied brusquely, dismissing the idea. Viacom, Freston continued, could also assist Facebook in extending its reach into an older audience. "I'm pretty much focused on high school and college," answered Zuckerberg. Why the two had flown all the way to New York somewhat mystified the Viacom executives. "It was a no-thank-you meeting," says one Viacom attendee. But Viacom did not give up.

In early November 2005, Michael Wolf, a longtime media industry consultant at McKinsey & Company, joined MTV as president, reporting to McGrath. He almost immediately took on the task of cultivating Zuckerberg.

Whenever MTV held a focus group among the college students who were its core demographic audience, they talked incessantly about Facebook. It gave Viacom a unique and early window into the power of this phenomenon. Viacom executives fretted that this new form of media might upstage them, and they wanted to get a foot in. Freston and his executives at MTV were also worried that News Corp.'s Fox networks were going to use their new ownership of MySpace to gain an advantage with TV advertisers. It seemed likely that a new kind of package deal including both social network and television components could soon emerge, at least for programming aimed at young people.

Wolf flew out to Palo Alto to visit Zuckerberg in his office. The Facebook CEO was in T-shirt and shorts, wearing his trademark rubber Adidas sandals. These had become so notorious that as Wolf arrived, an assistant was nailing one of the old worn-out sandals to a board. It was going to be given to one of Facebook's programmers as an award. Wolf considered himself to be cultivating Zuckerberg, and merely wanted with this meeting to start a cordial dialogue, but he did ask whether the CEO was thinking about selling the company. "I don't want to sell," Zuckerberg replied. "What kind of number might make you interested anyway?" Wolf asked. "I think it's worth at least $2 billion," said the kid who had launched Facebook in his dorm room twenty months earlier.

Shortly before this, an aggressive thirty-five-year-old dealmaker from Amazon.com named Owen Van Natta had joined Facebook as vice president of business development. The upbeat veteran executive was hungry for impact and authority, and had an enormous amount of energy. After only five weeks Zuckerberg promoted him to chief operating officer. Van Natta created Facebook's first strategic plan and immediately started bringing some order to what remained a chaotic and ragtag operation. The new COO wasn't shy about exercising his authority, and fired a number of engineers and other employees who had been recruited pell-mell earlier in the year. But Van Natta's greatest skill, honed at Amazon, was in negotiating deals. He would soon get a chance to prove himself.

Van Natta was annoyed that MTV's Wolf had figured out that the best way to reach Zuckerberg was by instant-messaging him and had thus been able to make an appointment directly with the CEO. Van Natta told Wolf in the future to go through him instead. Wolf ignored the instructions. Instead he periodically IM'd Zuckerberg to say that he planned to be in Palo Alto—whether or not it was true—and suggest a dinner. If Zuckerberg agreed, he'd fly out.

Wolf was only one of many top media and technology executives pursuing Zuckerberg. Facebook was hot. The office and the University Cafe down the block—the favored rendezvous—became a parade of big names. "The guys from NBC are coming by this afternoon." "When is that meeting with Microsoft?" "Peter Chernin is here!" (He was Murdoch's top deputy at News Corp.) "Did you hear that Zuck met with Dan Rosensweig from Yahoo?" There were meetings about a deal with

AOL, which owned the AIM instant-messaging system that Zuckerberg (and most of Facebook's users) used every day. For a while discussions centered on whether there was a way to build a special version of AIM for Facebook. Finally the companies struck a deal that enabled AIM members to invite their IM buddies to join Facebook. It quickly became a major source of referrals.

There was a lot of grumbling about all the Zuckerberg meetings, especially among the growing number of not-twenty-one-year-old executives whom recruiter Reed was helping hire. It appeared to many of these guys (they were almost all men) that Zuckerberg was willing to talk to anybody anytime about anything, especially if that person was a CEO. What did all these meetings mean? Was Zuck about to sell the company? Will we become part of Viacom or Yahoo or News Corp.? Are we all going to get rich? And for the younger, more idealistic ones— is this the end of the Facebook miracle? They sometimes wistfully discussed whether they ought to be looking for a new CEO.

Zuckerberg wasn't bothering to explain his thinking. He thought of these meetings as a learning process and didn't feel he had much to explain to the staff. After all, he had no intention of selling his company. And ironically, part of the problem stemmed from his good manners. He readily agreed, out of both curiosity and politeness, to meet with the honchos who came calling. And he listened politely, if impassively, during the day when Van Natta and the other older staffers were bending his ear. But late at night he continued to huddle more honestly with confreres Cohler, D'Angelo, Moskovitz, and frequently still with Parker. But so circumspect is Zuckerberg that sometimes they too were in the dark about his ultimate intentions. And everybody was painfully aware that he was in complete control of the company's destiny.

Reed was getting frustrated. She had helped bring in most of the older men who were feeling sidelined and who were now getting worried. She was proud of the quality of the team she'd helped assemble but saw the staff being overwhelmed with what she calls "dorm-room misinformation." Sean Parker may not have been an ideal company president, but he was pretty good at communicating. After Parker left, Zuckerberg gained more authority, but he didn't necessarily want it.

Reed had never gotten along with Parker, but it was almost worse with-out him. Communication seemed to be completely breaking down.

Politics were also getting heavy. Doug Hirsch, Facebook's vice pres-ident for product and a Yahoo veteran, was offending some of the other executives, many of them also newly hired. They thought he was trying to lead talks with the many companies that wanted various kinds of deals. Why was he not just sticking to product issues, they complained? Part of the reason was simply that Hirsch already knew many of the play-ers from his Yahoo days. They would call him up and suggest explor-atory meetings. Hirsch wasn't getting along very well with Zuckerberg, either. Hirsch was hired because so many people had told Zuckerberg that he needed someone else to head product development so he could focus on corporate matters. The CEO had been ambivalent all along about hiring someone as VP for product, since he considered that his own bailiwick. "Doug kind of felt he was there to be adult supervision," says Cohler, "and that was certainly not what any of us were thinking when we hired him." Hirsch himself says that some of the people he spoke to before he was hired had led him to believe he might eventually be a candidate to become CEO.

Reed had become a close observer of all the unhappiness, partly because she had one of the only private offices at the company, which she needed for candidate interviews. A copy machine had been moved out of an oversize closet and she had installed a *noren* Japanese door curtain for privacy. Next to her desk was a large sculpture of the Hindu god Ganesh, the remover of obstacles. But Ganesh didn't appear to be working. Many employees came to her office to gripe. Zuckerberg wouldn't listen, they said. Zuckerberg should be replaced. Zuckerberg didn't know what he wanted to do with the company.

Finally, Reed reached the end of her rope. "The morale of the ex-ecutives was imploding," says Reed. "The rumor mill was churning, and Mark wasn't communicating with anybody about what was really hap-pening. The team was almost ready to mutiny." Zuckerberg was on the East Coast at one of his many meetings. She decided to intercept him before he went back to the office. She instant-messaged him, asking to meet on his way home from the San Francisco airport. But his plane was delayed, and they didn't finally get together until 2:30 A.M. Reed came down from her home in Marin County, over the Golden Gate Bridge, and they converged in downtown San Francisco. Zuckerberg arrived in a stretch limo, which somebody had mistakenly ordered for him.

They sat in the neon glow of an all-night diner. Reed unleashed her frustration. "Mark—we've pulled together a team of thoroughbreds but they're locked in their stalls. Nobody knows what's going on. If you want to sell your company, then stop dicking around and say you want a billion dollars. Owen can go and get that offer. If it's two billion, say that. If you don't want to sell, then say that!"

"I don't want to sell the company," Zuckerberg answered, in his typical unflappable manner.

"Then stop taking all these meetings with Viacom and Time Warner and News Corp.! You're sending the wrong message." Then she unleashed her final barrage. "You'd better take CEO lessons, or this isn't going to work out for you!"

"So now you're finally being straight with me," Zuckerberg replied, turning more animated. "This is the first time I feel like you're telling me what you really think."

He had defanged her. She could no longer be angry. He actually was listening.

Over the next few weeks, Reed noticed a distinct change in Zuckerberg. For one thing, he did agree to start seeing an executive coach to get lessons on how to be an effective leader. He started having more one-on-one meetings with his senior executives. The week after the confrontation he called the entire staff together for Facebook's first "all-hands" meeting. He was feeling sick so he conducted the entire meeting sitting cross-legged on the floor.

Zuckerberg took the executive team to an off-site meeting where they could talk about goals and establish better communication channels. When Moskovitz heard about this he was dubious. "Do I have to fall back into people's hands so they catch me or something?" he said. "Because I'm not doing that shit."

Zuckerberg started doing a better job explaining where he thought the company was going. He wanted to make Facebook into a major force on the Internet, and not see it taken over by someone else, he repeated endlessly. He was getting better at explaining his priorities to his staff. His presentations included the simplest of slides—sometimes with just one bullet point, like "Company goal: grow site usage." The team was mollified. Reed cheered up.

Zuckerberg called Doug Hirsch into his office and they agreed it wasn't working. Hirsch wasn't officially fired, but it made no sense to stay. He'd been at the company four months. Zuckerberg had chafed at some of Hirsch's product initiatives, and they disagreed about some key projects Zuckerberg was planning. Hirsch was also aggressively coming up with ways to use Facebook's product to create more revenue, which in most companies would be routine. But in this one, back then, it was near apostasy. And there was talk among employees about unauthorized meetings he had supposedly had with solicitous companies like Google. To this day, many of Zuckerberg's young allies insist that Hirsch "was trying to secretly sell the company." But of course he couldn't have done that.

From the perspective of Moskovitz, who observed it all up close, this was just another example of a repeating pattern. "It's the same story with many of the executives," he says matter-of-factly. "Mark wanted to build the product out and focus on revenue as late as possible. And they wanted to make sure they had a business."

Reed brought needed structure to Facebook's management, but very few of the staffers she hired worked out in the long run. Hirsch was just the first of many who left within a year. Those in Zuckerberg's circle blamed her for bringing in people who didn't appreciate Facebook's unique mission and culture. Some of these stalwarts—the ones who survived—took to calling Zuckerberg's new executive coach "Wormtongue," after an evil adviser to the king in Tolkien's *Lord of the Rings*. Criticism was coming from the outside as well. Tech industry bloggers pointed to Facebook's revolving-door management and said it suggested internal chaos. But Zuckerberg's adviser Marc Andreessen gives the CEO credit for being decisive about making changes when people weren't working. There's no way, Andreessen says, that a fast-growing company can consistently make the right hiring decisions. Better to quickly remedy the inevitable wrong ones.

Zuckerberg preferred working with people his own age. He believed they were superior programmers, for one thing. Sometime later, at a small conference, he showed his stripes in talking to a bunch of other entrepreneurs. "I want to stress the importance of being young and technical," he said, according to the VentureBeat blog. "Young people are just smarter. Why are most chess masters under 30?" You can imagine how reading that made the growing number of Facebook executives in their thirties and forties feel.

Even as he was trying to learn better how to exert authority and manage his troops, Zuckerberg wasn't managing his own health very well. Or maybe the stress of Facebook was finally getting to him. He began fainting regularly, in the office and elsewhere, sometimes in the middle of a meeting or while he was sitting at his computer. His friends told him he should get more sleep and actually eat regular meals.

At a *Fortune* magazine dinner in early December 2005 that I hosted as the program director of a tech-centric conference called Brainstorm, I asked everyone at the large table to briefly talk about what was most on their minds. When his turn came, Jeremy Philips, a top strategist at News Corp. and close adviser to Rupert Murdoch, said how pleased his company was to have bought MySpace and mentioned that Facebook also seemed very interesting.

Viacom's Michael Wolf left the dinner in a panic. "Oh my God, they're talking to Facebook," he fretted. Viacom chairman Sumner Redstone would hit the roof if he lost out to Murdoch yet again. Wolf immediately called Zuckerberg and asked him point-blank if Facebook was considering selling to News Corp. Zuckerberg conceded that the two companies had talked, but said that he thought News Corp. was too Hollywood, and in any case media companies like that didn't understand technology ones like Facebook. Wolf didn't take the message as it was probably intended—to suggest that Viacom was also not appealing to Zuckerberg.

In mid-December, Wolf got in touch with a better offer than a meal at a local restaurant. He was planning to be in San Francisco with the Viacom corporate jet, he claimed. Would Mark like a ride back to New York for the holidays?

Zuckerberg took Wolf's bait. Since Viacom's corporate planes were in fact unavailable, Wolf chartered an unmarked, top-of-the-line Gulfstream G5 for the trip from San Francisco airport to Westchester County Airport, near Zuckerberg's parents' home in Dobbs Ferry, New York. Wolf flew out that morning from New York on American Airlines. The MTV executive was waiting aboard the G5 as if it were the most normal thing in the world when Zuckerberg arrived, late, about 5:30 P.M. Then, as Wolf had shrewdly planned, they spent five uninterrupted hours together aboard the plane. He was resolved to find a way for Viacom to buy Facebook.

For much of the trip, however, Zuckerberg was in control of the conversation. Zuckerberg interrogated Wolf about MTV's business. How did companies like Viacom make their money? How much did MTV charge for advertising? What amount of that was profit? How do you build your audience? Wolf tried to steer the conversation back to how MTV could work with Facebook. He talked about how MTV's ad sales team could use its access to big advertisers to help sell Facebook ads. And he noted that MTV's big hits like *Laguna Beach* and *The Hills*, watched by millions of teenagers and young adults, were perfect places to promote Facebook. Zuckerberg said he had noticed that during the hours those shows aired, Facebook's traffic slowed discernably.

During the trip Zuckerberg took to admiring the G5. "This plane is amazing," he said.

"Maybe you should just sell a piece of the company to us," Wolf replied. "Then you can have one for yourself."

Wolf invited him to sit in the jump seat in the cockpit as the powerful corporate jet landed at Westchester. When it pulled up to the private aviation terminal, two cars were waiting. One was Wolf's corporate black car to drive him into the city. The other was the Zuckerberg family minivan, from which Mark's parents emerged. They beamed and gave their son a big hug. It was as if he were merely coming home from a semester at college.

. . .

Wolf flew out to Palo Alto again in January 2006 and brought MTV's head of advertising strategy. Zuckerberg suggested they dine at the Village Pub in Woodside, the same fancy restaurant where he'd had his fateful dinner with Jim Breyer. He brought along Cohler and Van Natta. Wolf had an elaborate PowerPoint presentation showing how the two companies could work together. At the table he suggested a deal in which Viacom would buy a piece of Facebook in conjunction with a big ad partnership. Zuckerberg listened politely, but made it clear that he wouldn't even contemplate any deal that could involve him losing his absolute control over company decision making.

In early February, Wolf made yet another trip to Palo Alto. He and Zuckerberg were becoming chums, and they took a long walk around the palmy, well-groomed streets. For some reason they stopped by Zuckerberg's modest one-bedroom apartment. The place was messy, though mostly devoid of furnishings. There was a mattress on the floor with sheets askew, piles of books, a bamboo mat on the floor, and a lamp. Then they headed for dinner at a nearby restaurant. Wolf popped the same question he'd asked on the plane. "Why don't you just sell to us?" he asked. "You'd be very wealthy."

"You just saw my apartment," Zuckerberg replied. "I don't really need any money. And anyway, I don't think I'm ever going to have an idea this good again."

The conversation wove back and forth and Zuckerberg reiterated that he believed Facebook was worth $2 billion and wouldn't talk about selling it for less. "It wasn't like 'I want $2 billion,'" says Wolf. "It was 'If you pay me $2 billion I don't want to sell. Thank you.'" Zuckerberg finally said it made more sense for them to just talk about some kind of partnership.

A thwarted Wolf went back to New York and met with McGrath and Freston. They weren't interested in a partnership. They—and Redstone—wanted very badly to own Facebook. So Freston decided to just make an offer. He sent Zuckerberg a letter proposing Viacom would pay $1.5 billion to buy the two-year-old company. The money was to be paid out 51 percent in cash and the rest over time, depending on how well Facebook's business performed. It was by far the most significant and concrete offer Facebook had ever received. Zuckerberg didn't even respond.

A week or so later, Wolf called Zuckerberg and they had a few desultory conversations, to no effect. Wolf met with both Peter Thiel and Jim Breyer, complaining of Zuckerberg's tepid reaction, but both said they couldn't do much to intervene. Van Natta, by contrast, confided to Wolf that he himself was trying to convince Zuckerberg to sell.

Meanwhile the Viacom team had heard that Yahoo might be talking to Facebook as well. The elite precincts of media mogulhood are like a small town. It happened that Viacom's Freston played tennis regularly in Los Angeles with Yahoo CEO Terry Semel. One day on the court Freston tried to sound out Semel to learn whether he was talking to Facebook. He got the impression the answer was yes. The pressure on Viacom increased.

Up to this point, Wolf had taken painstaking measures to keep the Facebook talks secret. Few others at Viacom even knew about them. Freston and McGrath thought one reason Murdoch had been able to swoop in and bag MySpace was that Viacom's talks with that social network had been too public. But at the end of March, *BusinessWeek's* online edition published a story titled "Facebook's on the Block." It reported the incomplete fact that the company had turned down $750 million dollars and hoped to get $2 billion. The article didn't say Viacom had been the bidder, but speculated about its interest. To Wolf and his colleagues, this was embarrassing. They presumed Facebook had leaked the info to elicit additional bids. And sure enough, shortly after the article appeared, Zuckerberg called and said he still wanted to talk.

Then Facebook came close to selling out. Van Natta and Zuckerberg came to New York. Wolf flew back out to Palo Alto. With several Viacom colleagues, Wolf camped out in a Facebook conference room. Zuckerberg, Cohler, and Van Natta would come in, negotiate, then retreat to another nearby conference room. The Viacom team would walk around the block. Back into the conference room. Another tête-à-tête. Mark wanted more cash up front. Viacom wanted guarantees on performance before it would pay the remainder of the $1.5 billion. Van Natta wanted fewer restrictions on the payout. Wolf finally agreed to increase his initial payment to $800 million in cash. But they continued to quibble about the remaining $700 million. Neither side had an investment banker assisting them, as would be routine in most such talks. Wolf knew Zuckerberg well enough to realize that bringing in cold-blooded Wall Street experts would only further spook him.

But Wolf's bargaining leverage was limited. Viacom's chief financial officer was leery of paying too much for a company that for all its online presence remained puny in financial terms. Facebook had only seen around $20 million in revenues over its entire history to that point, with effectively no profit whatsoever. It planned on $22 million in revenue for 2006 and $55 million for 2007, executives told Wolf, but the Viacom delegation was skeptical it would reach those numbers. Paying $800 million was really stretching it.

In the end the two sides could not agree on how Facebook would earn its additional $700 million. The Facebook negotiators felt the deal's terms were too complicated and the payout uncertain. Zuckerberg seemed to be getting cold feet anyway. He was saying things like "Google was smart not to sell early. Look at how well they did." Wolf responded that Google had hundreds of millions of dollars of profits before it went public, and Facebook had none. But to Zuckerberg, what was more significant was that Facebook had by then become the seventh-most-trafficked site on the Internet, with 5.5 billion page views in February, according to the measurement firm comScore Media Metrix.

As the Viacom deal was petering out, Facebook did some of its own financial maneuvering. It raised more money from venture capitalists, but for this second VC round (known as Series C because it was the company's third financing), the pre-investment valuation was $500 million, five times the $97 million postinvestment valuation Accel had agreed to eleven months earlier. Premiere venture firm Greylock Partners led the April round, joined by Meritech Capital Partners. In addition, Peter Thiel and Accel Partners each put in more money and added to their Facebook holdings. Altogether Facebook received an infusion of $27.5

million. It significantly relieved the financial pressure and made it considerably easier for Zuckerberg to walk away from Viacom.

Facebook's success was attracting another sort of attention—from international imitators. Though the company had begun expanding to select elite schools in English-speaking countries outside the United States, it had no presence in Asia and virtually none in Europe. A site called studiVZ (from the German for "student directory") in Germany now borrowed Facebook's design, making red the elements that on Facebook were blue. Otherwise it was a pretty shameless imitation. It launched at German universities in October 2005 and was an instant success. By January 2007 it had 1.5 million users and sold to the powerful Holtzbrinck publishing firm. Facebook was so worried that this might preclude its ultimate success in Germany that in late 2007 it came close to buying studiVZ—for about 4 percent of Facebook's total equity. The prospect of a purchase was made easier, ironically, because the imitation was so complete. That would make integrating the two services much easier. Another imitator, which launched around the same time in China, called Xiaonei, blatantly copied some of Facebook's software code and even initially included at the bottom of each page "A Mark Zuckerberg Production." Xiaonei too was a hit, garnering many millions of users.

Despite his MySpace coup, News Corp.'s Murdoch grew ever more intrigued by Facebook. He and Zuckerberg got to be fairly good friends. The mogul was charmed by the passion of the young CEO, and Zuckerberg liked Murdoch's big-picture view of how media was changing. Murdoch, almost uniquely among media leaders, had accepted that the Internet was transforming the landscape for all media companies. He considered his purchase of MySpace just one in a series of major moves. But he couldn't understand why Zuckerberg thought Facebook, which had far fewer users at that point, was worth several times what he'd paid for MySpace. The conversations never got as serious as those with Viacom, but they would gain momentum, then peter off as Zuckerberg's interest declined.

Zuckerberg was getting a little cocky. Everybody wanted to talk to him. Every company seemed to want to buy Facebook, and everybody seemed to want to use it. And he had noticed another thing—every offer he got for the company was higher than the last. Meanwhile, the service's growth was steady. If it was going to keep getting bigger, it would keep getting more valuable. He didn't want to sell anyway, so there was no urgency to any of these conversations.

But Facebook was still burning tons of cash. It couldn't keep endlessly pulling in investment money to cover its losses, no matter how much contempt Zuckerberg had for ads. Luckily, Google, Microsoft, and Yahoo all wanted to talk about a deal to place display ads on Facebook. Zuckerberg authorized his deputies to begin negotiations. To him it seemed like easy money. He wasn't going to give them much onscreen real estate anyway.

Facebook was now so successful it was beginning to saturate the college market, operating at thousands of schools. At almost every school it opened, the majority of students became users. Its success at high schools reinforced Zuckerberg's belief that Facebook had the ability to spread quickly among new groups. What mattered was that the target group needed to include lots of dense, overlapping relationships.

And what was the mother of all such communities? The workplace. Zuckerberg decided to launch what he called work networks. It would be Facebook's first effort to recruit adults to its service. A work network would be set up at a company the same way Facebook established a closed student network at each university. The default privacy setting was that members of such a community could all see one another's information. Zuckerberg believed work networks would extend the company's ubiquity out of the academy into the whole country, and maybe even ultimately the world, or at least to everyone who worked for companies. It was very different from Facebook investor Reid Hoffman's LinkedIn, which was structured more as a résumé-based network and did not so much emphasize day-to-day communication or workplace social connections.

In May 2006 the work networks debuted, but not much happened. The world barely noticed. Facebook created networks for a number of companies, opening the doors, but few passed through. One exception was the unique workplaces of the U.S. Army, Navy, and Air Force. The intensity of shared experience among young people in the military is apparently much like that in college. Facebook made sense there. But in most of the big companies where Facebook initially set up networks, there was little if any response from employees.

Few in business knew that Facebook was opening up. And Facebook was developing a bad reputation. At almost exactly the same time work networks debuted, the *New Yorker* published a lengthy profile of Zuckerberg and Facebook, the most in-depth coverage the company had ever received. Author John Cassidy made the site seem like a curiosity, focused a good bit on the Winkelvosses' lawsuit, and implied that Facebook's users were antisocial. "Clearly one of the reasons the site is so popular is that it enables users to forgo the exertion that real relationships entail," he wrote. He also quoted a sociologist who speculated that the main reason Facebook was so popular was "voyeurism and exhibitionism."

Facebook still seemed to nonusers to be mostly about dating and doing pointless, possibly suspicious things like poking people. Whenever you added a new friend on Facebook back then, a box popped up that asked you how you knew them. One of the options was "we hooked up." How could this be a service for professionals? And Facebook faced a chicken-and-egg problem. Adults didn't want to join until other adults were already there.

Perhaps Facebook only worked for students after all. Maybe adults didn't need this kind of service, many Facebook executives worried. The mood around the office darkened. Even though growth was continuing strongly among college and high school kids, if adults didn't want to join Facebook then perhaps something was wrong with Zuckerberg's theories. He was disappointed and befuddled. This was a major setback. Maybe the world wasn't becoming more transparent as quickly as he had thought. "It was the most wrong he'd ever been at Facebook," says Cohler, "and the first time he'd ever been wrong in a big way."

Zuckerberg had other big changes in mind, but if adults weren't going to respond to Facebook some of the changes would fail. As summer began, Facebook's board debated how serious the problem might be. David Sze, who had spearheaded Greylock's recent investment in Facebook and was an official board observer, found himself having to reassure its members. At one meeting Moskovitz, who also attended as an observer, quizzed Sze about whether he regretted investing, given the unexpected difficulties with work networks. At that moment Sze was more optimistic than Facebook's otherwise chronically upbeat board.

That summer of 2006, for the third year, the company rented a Facebook house, occupied mostly by recent arrivals in Palo Alto. One of Facebook's lawyers argued that a company house was too great a legal liability, but Zuckerberg overruled him. The CEO decided the company should pay half the rent because anyone should be able to come by and use the pool. In fact the pool was only used sparingly, since its heater was broken and the water was always around 100 degrees.

Zuckerberg kept a room for himself in the house to use on weekends but the rest of the time lived separately in his own apartment. He had split up with the Berkeley undergraduate he'd been dating and reunited with his old girlfriend, Priscilla Chan, who he had met while in line for the bathroom at a Harvard party. She had graduated with his original Harvard class of 2006. Instead of a diploma, Zuckerberg had a company worth over $500 million and with almost a hundred employees. After some negotiation, Zuckerberg arrived at a deal with an insistent Chan: while they wouldn't live together they would spend a minimum of 100 minutes of time alone each week and have at least one date, which would not be either at his apartment or at Facebook.

But the company retained a collegiate air. Employees called another nearby house "the frat house." Nine people lived in its four bedrooms, many of them recently arrived Harvard-trained programmers. In the window were three big Greek letters that had originally decorated the first Facebook office on Emerson Street—Tau Phi Beta—for The Face Book.

It's not hard to understand why the lawyer had concerns. The dining room was turned over to Beirut beer-pong tournaments, but one resident, Chris Putnam, was only nineteen. As a sophomore at Georgia Southern University, he successfully hacked his way into the company's servers and made two thousand Facebook profiles look like they were on MySpace. He inserted a note in the code saying he didn't intend any harm. The episode impressed Zuckerberg and Moskovitz so much that they hired him.

At the frat house employee recreation could also be more productive. "People would just come over and code, or party and watch *Lost*," reports programmer Dave Fetterman. "We could still fit everyone in the company into the house for a party. At night we'd have beers, watch TV, think up new ideas and just start coding them right there, either in someone's room or out in the yard. Mark or Dustin would show up. They were usually the first ones to open up their laptops." The programmers would sometimes combine partying and working at what they called "push parties." They'd load new software onto the site and "push" it live from right there at the frat house.

Big advertisers were beginning to experiment cautiously with Facebook. This wasn't just little record companies advertising Gwen Stefani songs anymore. The giants of marketing now were getting interested in Facebook. But it was a different environment than they were used to. The company's still-small group of ad salespeople pushed clients to craft messages and offerings unique to the service, in keeping with Zuckerberg's near contempt for traditional advertising. (When he hired new ad sales boss Mike Murphy in March 2006, Zuckerberg told him, "I don't hate all advertising. I just hate advertising that stinks.") Even COO Van Natta, a hard-charging industry veteran, had swallowed hard and accepted Zuckerberg's dictate that advertising should always be useful for the user. Though his mandate was revenue, he had taken to saying things like "We almost shouldn't be making money off of it if it isn't adding value."

Chase credit cards was an important pioneer. Working with a small New York ad agency called Noise Marketing, it created the Chase +1 card, specially designed for college students and only available to Facebook users. The card was black, because that's what students said they wanted. It offered something Chase called Karma points, which you could redeem for modest rewards like concert tickets. But unlike most rewards cards, you could accumulate points without spending large amounts. That made sense for students because they typically make only small purchases. Each purchase, no matter how small, garnered twenty points. You also got points for joining Chase's sponsored group on Facebook, as well as for taking an online course on how to manage your credit. And Chase made its card "social." You could give your Karma points to your Facebook friends.

A week after the program launched, 34,000 students had already joined the group, and Chase soon issued thousands of cards. The bankers were pleased, and Facebook had taken an important step toward proving that customized advertising could work.

A few months later, Procter & Gamble tried something similar. Its CEO, A. G. Lafley, had begun talking about the need for P&G to get closer to its consumers. After reading about this, Facebook ad salesman Colleran did one of his masterful cold calls to find out if P&G was targeting any of its brands at the college market. It turned out that while P&G's Crest White Strips teeth-whitening product had never been aimed specifically at college students, company data showed that the strips sold particularly well at Wal-Marts located near campuses. Colleran and P&G marketers came up with a Facebook campaign called Smile State.

Much as Chase and Apple had done, P&G created a sponsored group on Facebook for Crest White Strips. It advertised the Smile State group only to users who were students at one of twenty large state universities located near Wal-Marts. Any student who joined got tickets to an upcoming college-oriented Matthew McConaughey movie called *We Are Marshall*. In addition, the schools that enrolled the most members in the Crest White Strips group got a concert organized by Def Jam Records. Over 20,000 people joined. To have 20,000 people explicitly expressing affinity for Crest White Strips using their real name is the kind of thing that gives marketers goose bumps. It was a huge win for P&G and for Facebook.

Zuckerberg remained uninterested in advertising that interrupted the Facebook experience and distracted users' attention, no matter how lucrative it might be. In May 2006, Sprite was relaunched with new packaging and a tongue-in-cheek ad campaign aimed at young people that was meant to be brash and obvious. The soft drink's ad agency offered to pay $1 million for a banner ad that would turn Facebook's entire home page green for one day. Zuckerberg didn't even consider taking the money. Nor was the CEO interested in impressing people to get their business. The first time the top executive of a big San Francisco digital ad agency visited Facebook, he ran into Zuckerberg, who was barefoot and wearing NBA basketball shorts that hung below his knees.

Most advertisers were still uncertain what exactly Facebook was, not to mention how to take advantage of it. But in June the world's third-largest ad agency declared itself in Facebook's camp with a dramatic gesture. The Interpublic Group committed to spend $10 million on Facebook ads over the coming year on behalf of its clients. As part of the deal the ad giant also bought half a percent of Facebook's stock. "Young and tech-savvy consumers are increasingly shunning traditional media vehicles and defining themselves and their community online," Interpublic CEO Michael Roth said in a statement. He also noted that 65 percent of all U.S. college students by now were maintaining a Facebook profile.

In August, Facebook got another major acknowledgment—this time from a titan of the tech industry. First, MySpace announced a major three-year, $900 million deal with Google to operate a search function inside MySpace and to place advertising there. It was such a large deal that by itself it turned Murdoch's investment in MySpace profitable. It was the second time a huge MySpace transaction shed reflected glory onto Facebook. The first time—when Murdoch bought it—it had made Facebook look valuable. This time it made Facebook ad inventory look like a gold mine.

COO Van Natta and newly hired Vice President of Business Development Dan Rose, whom Van Natta had hired from Amazon, had already begun talking with the companies with the biggest online display ad operations—Google, Microsoft, and Yahoo. Facebook already had a small deal with Microsoft's MSN online division to sell ad space.

Nothing motivates Microsoft like the desire to best Google. A day or so after the MySpace-Google deal was announced, Rose called up Microsoft, knowing the software colossus had battled for the MySpace deal and lost.

Rose immediately got a positive reaction to his inquiry. Yes, the Microsoft executive he spoke to said, he would love to talk about a similar deal with Facebook. "What are you asking for?" he asked. Van Natta and Rose huddled and quickly came up with what they thought would be a juicy deal. They proposed that Microsoft use its ad sales network to represent Facebook's banner ad inventory and guarantee a certain CPM for every ad it placed. They didn't even get an argument. "Okay, we'll be down there tomorrow to iron it out," said Rose's eager Microsoft counterpart. It took some work to finalize the details. Says one Microsoft negotiator: "Mark was adamant about preserving the user experience and the layout. It drove our ad people crazy because it made it very hard to deliver standard Internet ad units."

It was a transformative deal. Facebook now had a large and lucrative new revenue stream. Instantly Microsoft turned 2006 from another money-losing year for Facebook into a highly profitable one. A few months earlier, Viacom's Wolf had been shown internal projections aiming at $22 million in 2006 revenue, but Facebook ended up at least doubling that. Microsoft's payments accounted for well over half of company revenue for 2006. For 2007, the Microsoft deal guaranteed Facebook $100 million in revenue.

Perhaps Zuckerberg's CEO lessons were paying off. He was letting the experienced Van Natta play a role not unlike the one Parker had played earlier—serving as Mr. Outside and building the business, letting Mark focus on improving Facebook's product. Van Natta was managing bigger and bigger deals with partners like Interpublic and Microsoft. The executive team—purged of some of Robin Reed's hires—was coalescing. Though her in-house recruiting stint had been extended more than once, by now she was gone. The team didn't want to admit it, but she had helped the company grow up.

Viacom had abandoned trying to buy Facebook, but his talks with Michael Wolf had taught Zuckerberg a lot about deals and about how the media industry works. That would serve him well in the coming years. And inside the company he seemed more like a leader.

9

2006 "I can't find out what's going on with my friends!"

The astonishing success of Facebook's photos application led to a bout of soul-searching at the company. What was it, Zuckerberg and his colleagues asked themselves, that made photos so successful? Well, one thing was that you could so easily find new photos your friends uploaded. Each person's profile included a "dashboard page" that showed which photo albums had most recently been updated. It seemed that users wanted to know what was new. Another recent innovation had been to order the list of friends on each user's home page according to which profiles had been changed the most recently. They called that "timesorting," and it won raves from users. Each time someone changed their profile picture, it quickly led to an average of twenty-five new page views.

What people did on Facebook was look at other people's information. They were eager to learn what was new, what had changed, what had happened that they didn't already know. Studying your friends' profiles was an obsessive activity, but not a very efficient one. Click through and try to figure out whether anything had changed since the last time you visited. Was he still single? Did this photo mean she'd been to the Caribbean? How come he went to that party and didn't tell me? Click click click. The information was good—you wanted to know it—but it was tedious to find.

So the company's young leaders came up with the idea to build a page that showed not just the latest photos your friends had added, but all the things that had recently changed on the profiles of your friends. "We started asking, 'How do we get people the information they most care about?'" says Moskovitz. "We wanted to build a screen that showed everything. So we came up with the idea for the News Feed."

The new tool they arrived at would help users find the information that most mattered to them at any given moment. That might include everything from which party a friend planned to go to on Friday to updates about the political situation in Tajikistan someone might have posted as a Web link. The point was to make sure you saw what you cared about, whatever it might be. The order in which information would be presented would depend on what you had shown—by your behavior—you liked to look at. Zuckerberg explained it to colleagues: "A squirrel dying in front of your house may be more relevant to your interests right now than people dying in Africa."

All this brainstorming took place in the early fall of 2005. Shortly afterward, Adam D'Angelo talked to a new hire, Chris Cox, about building the News Feed. "I saw a glimmer in his eyes," says Cox. "I could tell that for him it wasn't about wanting to make money. He said, 'Look, this is such a broken problem—I can't find out what's going on with my friends!' The Internet could help you answer a million questions, but not the most important one, the one you wake up with every day— 'How are the people doing that I care about?'"

They set to work on the News Feed. "For the next eight months, it was our labor of love," says Cox, a tall, laconic, and brainy Stanford grad who had studied computer science, psychology, and linguistics. The idea was audaciously ambitious: to write a set of software algorithms to dissect the information being produced by Facebook's users, select the actions and profile changes that would be most interesting to their friends, and then present them to those friends in reverse chronological order. Each person's home page would thus be completely different, depending on who their friends were. "It was the biggest technology challenge the company had ever faced," says Sean Parker.

The average user of Facebook at that time had about 100 friends. The software would have to watch every action generated by every one of those people. Then, each time you went onto the service, it would rank the activity of all your friends based on the likelihood that you would see it as interesting. That calculation would be based on, among other things, your previous behavior. Perhaps you noted that you were feeling glum or that you were going to the movies, or you uploaded a photo, indicated you like the new Wilco album, or posted a link to a segment of the *Daily Show*. Facebook's software would detect this new information and decide whether to send it to your friends, based on what it calculated is likely to interest them. It would infer this based on its observations of your friends' previous behavior. If they liked hip-hop they might not get the Wilco info. If they never watched videos they might not see the *Daily Show* link. It would apply such logic to every sort of information and activity on the site. It would repeat this process every fifteen minutes or so. Now multiply all this by 6 million—the number of active users Facebook had at the project's outset. This was a massive engineering and product design challenge.

The News Feed would be a radical change. "It's not a new feature, it's a major product evolution," said Zuckerberg at the time. It would remake Facebook. It was necessary as a foundation for future innovations he was already thinking about. He evangelized it with conviction to the company's engineers and product designers, not always successfully. "Many of us were 'No no no, we hate this!'" says product manager Naomi Gleit.

Though Zuckerberg remained opposed to selling the company, many in his orbit at Facebook felt it was just good business to learn what other potential buyers might pay. Owen Van Natta was expert at ferreting out offers. In the late spring of 2006, following the demise of the Viacom talks, which topped out at $800 million cash, Zuckerberg and the board concluded that if someone bid $1 billion cash for Facebook they would consider it seriously. Zuckerberg agreed partly because he was worried that the failure of the work networks might mean that his baby wasn't destined to be as big as he'd thought.

Meanwhile, a few towns south, in Sunnyvale, Yahoo's executives were worried. They saw social networking taking deeper and deeper hold even though they had no position there. CEO Terry Semel was becoming increasingly enamored of Facebook. Chief Operating Officer Dan Rosensweig had become a fan earlier and had gone out of his way to get acquainted with Mark Zuckerberg in 2005. On more than one occasion Rosensweig made it clear that Yahoo would talk about an acquisition if Zuckerberg were interested. He wasn't.

By June, Yahoo's executive team unanimously concluded they should buy Facebook. Semel approached Zuckerberg and they began talking. It quickly appeared possible that Yahoo might be willing to pay $1 billion. Semel, Rosensweig, and Yahoo Chief Strategy Officer Toby Coppel embarked on a series of negotiations with Van Natta, Cohler, and Zuckerberg, many of the meetings taking place at Van Natta's Palo Alto home. (The CEO's furniture-free one-bedroom apartment wasn't suitable.)

Zuckerberg didn't know whether to feel celebratory or defiant. To express a little bit of both he had one of his product managers buy $500 worth of illegal fireworks for an all-company July 4th party. Setting them off in a Palo Alto park led to an awkward but fleeting encounter with the police. The next day Yahoo formalized its offer in a term sheet Semel messengered to Zuckerberg.

From the perspective of the CEO and allies like Moskovitz, this development was jarring. Moskovitz, like Zuckerberg, had no real interest in selling. "The way it was pitched to me," he recalls, "was 'It would be irresponsible not to figure out what our valuation is. We're not trying to sell the company.' And that quickly snowballed into, 'Okay, now there are term sheets and we're going to have to pretend to talk.'"

Board member Breyer saw it differently. This was potentially a major opportunity for a lucrative "exit," to use the term VCs use when they make a lot of money. If the deal went through, Accel would make more than ten times its investment in just fourteen months. "I was calling for board meetings, calling for discussion," Breyer remembers. "I said, 'We have to document this, and go through a process where we talk about the pros and cons. You can't just dismiss this out of hand. We have a lot of employees who we're representing. This is real money to them.'" Of the company's young leaders, he says, "Once the offer came in, even though that had been our number, they didn't want to do it. Mark was definitely feeling at that point that he didn't want to sell. So there was absolutely tension."

At one board meeting, Zuckerberg lost his patience. "Hey Jim, we can't sell it if I don't want to sell it," he said bluntly.

"I know that, Mark," replied a piqued Breyer. "But we said our number was $1 billion. Let's go through the analysis."

Breyer was by no means the only person lobbying for a sale. Again, there were two camps, and it was mostly older employees versus younger ones. The relatively older Van Natta and Cohler (in their early thirties and late twenties respectively) both wanted to sell. Sean Parker, still a major shareholder, was allied with Zuckerberg and Moskovitz. He thought Facebook was just getting started. Peter Thiel, older but very sympathetic toward Zuckerberg, spent hours talking with the CEO about whether it made sense to sell. Thiel wanted Zuckerberg to consider it, but remained deferential to the founder. "In the end Peter was willing to support me," recalls Zuckerberg. "Jim pushed a bit harder. Pretty much everyone else wanted to sell the company."

Moskovitz, Zuckerberg's longest-standing partner, was one of the few others firmly opposed to selling. "I was sure the product would suffer in a big way if Yahoo bought us," Moskovitz says of that time. "And Sean was telling me that ninety percent of all mergers end in failure." He and Zuckerberg were also closely following the outcome of Google's acquisition in early May of Dodgeball, a company that used cell phones to help you track the physical location of your friends. "We saw that Dodgeball was going to shit," says Moskovitz. "And Google was the mecca of start-ups. If an acquisition there was going to fail I didn't feel great about going to a company that was known for being kind of behind the times."

Zuckerberg was certain that Yahoo's bid, impressive as it was, would be seen as way too low if the News Feed succeeded as he hoped it would. The launch was planned in less than two months, when the new school year began. And Facebook was also planning at almost the same time to make another dramatic change—it was going to open itself up so that anyone could join. No longer would it be necessary to be affiliated with a college, high school, or workplace network. This open registration came to be called "open reg." Unlike work networks, it wasn't simply taking the college model and applying it in a new market. It was a wholesale shift—declaring that Facebook ought to be for everyone. The company didn't drop its old structure. It still slotted every user into a network. But if you weren't in a school or a workplace, you could just join the network for your city. This would be the true test of Facebook's broader appeal beyond students.

Cohler and Breyer were both worried that the failure of work networks might mean that open reg would bomb as well, and that Facebook would never go beyond the student market. "We were saturated in college," says Cohler. "We were saturating in high school. MySpace was very strong in the twenty-something demographic. And Mark had what felt at the time like a blind belief that large-scale adoption of the product by adults was just going to work. He had always been right about those things and many of us had been wrong, up until the work networks."

If Facebook was not going to jump beyond colleges and high schools into the broader population, then its growth had almost certainly topped out. To Cohler that meant the Yahoo offer might be the best they'd ever see. "Mark, I'm open to having my mind changed," said Cohler. "Explain it to me."

"I can't really explain it," answered Zuckerberg. "I just know.'"

In the opinion of many of the company's more veteran employees and investors, Facebook had a golden opportunity to capitalize on its uniquely thorough penetration of the college market. Some said Facebook looked like MTV in its early years, when its rock-video network created a new form of media that young people simply couldn't stop watching. Those who held this view argued Facebook risked undermining its standing among high school and college kids by inviting a bunch of uncool adults into the service with them.

Zuckerberg disagreed. His view was consistent and clear—Facebook needed to go beyond college and become a site everybody could use to connect with their friends. He and Parker and Moskovitz had been saying since mid-2005 that Facebook was not meant to be cool, just useful. If younger people were turned off as the site broadened demographically, so be it. Zuckerberg knew that people on Facebook weren't very aware of anyone outside their own social circle anyway. Older people might join in droves without the average college kid even noticing.

The tension with Breyer and his executives and the gravity of the question of whether or not to sell to Yahoo gnawed at Zuckerberg. Some nights, unable to sleep, he would get into his car and just drive, with his Green Day and Weezer CDs cranked up loud. He spent hours pacing around the pool at the company house, trying to think things through. His girlfriend, Priscilla, lying on a nearby chaise one day, said to a friend, "I hope he doesn't sell it. I don't know what he'd do with himself." Zuckerberg had a talk around this time with his older sister Randi, who worked in marketing at Facebook. "He felt really conflicted," she recalls. "He said, 'This is a lot of money. This could be really life-changing for a lot of people who work for me. But we have so much more opportunity to change the world than this. I don't think I'd be doing right by anyone to take this money.'"

The negotiations at Van Natta's house continued for the first two weeks of July. Yahoo's lawyers conducted due diligence on the company's financials. Finally the two sides reached an agreement in principle for Yahoo to buy Facebook for $1 billion cash. But for all that, some on the Yahoo side could tell that Zuckerberg remained unconvinced. He seemed to be taking his sweet time at every phase of the talks. They weren't sure he was really willing, despite what might have been hammered out with Van Natta. They were right. And some of Zuckerberg's other attitudes frustrated the Yahoo team as well. For instance, one Yahoo negotiator recalls, "Mark had no interest at all in accommodating advertising in Facebook's product."

Then all the tension was relieved with unexpected suddenness. In mid-July, Yahoo announced second-quarter financial results. Wall Street viewed them as disappointing and knocked Yahoo's stock down 22 percent in a single day. Shortly thereafter, CEO Semel got cold feet, much as had Viacom's CFO earlier in the year. How would Wall Street react if Yahoo spent a huge amount on a company with so little revenue? Semel reduced his bid to $850 million, recognizing it could end the deal. It did. His deputy Rosensweig called and told Zuckerberg that Yahoo was reducing its $1 billion offer. As soon as he got off the phone, a grinning Zuckerberg strode over to Moskovitz's desk a few feet away and gave a big high-five. In a ten-minute conference call, Facebook's board rejected the offer. Even Breyer was comfortable with the decision.

As all this was under way, executives at other media and technology companies were starting to ask if they ought to buy Facebook. Rumors of Yahoo's billion-dollar bid were circulating.

At Time Warner, discussions about Facebook briefly turned serious. AOL CEO Jonathan Miller wanted to buy it. He saw community as the core of AOL, manifested in its chat rooms, forums, and AIM. Facebook would fit in perfectly, he thought. But AOL was just a division of Time Warner. Miller couldn't proceed without the concurrence of the parent company's leaders, who had turned down previous proposals he'd made for acquisitions. Miller also knew Zuckerberg would not want to take Time Warner's stock, much derided at the time for performing so poorly. Any deal would have to be for cash.

So Miller got creative. A partnership with another Time Warner division, he concluded, might help overcome corporate resistance. He succeeded in recruiting Ann Moore, CEO of Time Inc., the magazine division, for a possible joint bid for Facebook. The two concocted a plan by which each would sell assets to assemble cash for a Facebook purchase. AOL would sell MapQuest as well as its Tegic software, used on cell phones to predict words you're trying to key in. Miller hoped to get as much as $600 million altogether. For her part, Moore would sell Time Inc.'s British magazine publisher IPC for around $500 million. Then they'd have enough for a cash bid for Facebook.

But when they brought their proposal to Jeff Bewkes, Time Warner's president, he shot them down. He said if they could live without those properties they should go ahead and sell them, then turn the cash over to the parent company. If they wanted to do a Facebook acquisition later they should come to him and he would consider it. That was the end of that. Zuckerberg never even heard about the plan.

As the summer continued, excitement built inside the company about the twin launches planned for the first weeks of school. Facebook's News Feed team was putting on the finishing touches. And the people overseeing open registration had decided to also inaugurate a new way of getting friends to join you on the service. You would be able to download your email address book from any of the major email providers—Hotmail, Yahoo mail, Gmail, or AOL—and with a few clicks find out who in your address book was already on Facebook. You would also be able to send emails to anyone who wasn't on Facebook, inviting them to join. So central was this element that some began referring to open registration as "Address Book Importer."

Developing the News Feed was by far the most complex and lengthy project Facebook had ever tackled. But by midsummer a version was working. One night, sitting in his living room, Chris Cox saw the first News Feed "story." On his home page was one brief line: "Mark has added a photo." "It was like the Frankenstein moment when the finger moves," Cox marvels. The News Feed would eventually be comprised of a long list of such alerts customized for each user. The conceptual model for the News Feed was a newspaper that was custom-crafted and delivered to each user. Facebook called each little alert item a "story." The software that calculated which stories should go to each user was deemed "the publisher."

There was extraordinary anticipation at the company as News Feed's debut neared. Dave Morin, an employee at Apple, was being recruited by Parker and Moskovitz to join the company at that exact anxious moment. (Parker may have stopped getting a salary, but his passion for Facebook's success was unabated.) Morin recalls a conversation with Parker the night before News Feed launched. "Morin, tomorrow will be the day that decides whether or not Facebook becomes irrelevant or becomes bigger than Google," Parker intoned. Moskovitz had a less portentous thought for Morin. "Tomorrow you're going to love the new home page so much," he said, "you're going to want to work here for free!"

Facebook turned on News Feed in the wee hours of the morning of Tuesday, September 5. Everybody had been working so hard that the office was a wreck—wires and papers strewn everywhere. The corporate refrigerator was packed with cheap Korbel champagne for a big celebration. People pulled it out and began swigging directly from the bottles. Some people even brought in New Year's noisemakers. This was something to celebrate. As they pushed the button to officially turn the feed on, a crowd gathered around a monitor. Zuckerberg was there, barefoot, wearing a red T-shirt from New York's CBGB's nightclub and black baggy basketball shorts.

Ruchi Sanghvi, the News Feed product manager, posted an upbeat note on the Facebook blog, "Facebook Gets a Facelift." "We've added two cool features," she wrote guilelessly, "news feed, which appears on your homepage, and Mini-Feed, which appears in each person's pro-file. News feed highlights what's happening in your social circles on Facebook. It updates a personalized list of news stories throughout the day, so you'll know when Mark adds Britney Spears to his Favorites or when your crush is single again. . . . Mini-Feed is similar, except that it centers around one person. Each person's Mini-Feed shows what has changed recently in their profile and what content (notes, photos, etc.) they've added."

Now a user's home page was entirely composed of algorithmically selected snippets telling them what their friends were up to. Here are some examples that appeared in users' News Feeds: David Walt added new photos; Monica Setzer is now single; Amanda Valerio changed her profile picture; Alex Stedman left the group UCSB Students Against Beer Pong; Dan Stalman and Alex Rule are now friends; Lauren Chow is at-tending The Gods Must Be Crazy; Garrett Tubman is better cause zackie just cheered him up; and Updated: 14 of your friends joined the group Students Against Facebook news feed (Official Petition to Facebook).

Yes, there was a problem. Apparently Facebook's users hated News Feed. After the engineering team pushed the code live, they sat and watched as reactions from Facebook's 9.4 million users started coming in. The very first one read, "Turn this shit off!" Photos of the evening show a celebration suddenly turned sour, as slightly inebriated staffers stopped gleefully brandishing their Korbel and began glaring at screens suddenly filled with cascading complaints.

Thus began the biggest crisis Facebook has ever faced. Only one in one hundred messages to Facebook about News Feed was positive. At Northwestern University in Illinois, a junior named Ben Parr woke up Tuesday morning, logged into Facebook, and did not like what he saw. He quickly created the anti–News Feed group "Students Against

Facebook news feed." "You went a bit too far this time, Facebook," he wrote. "Very few of us want everyone automatically knowing what we update . . . news feed is just too creepy, too stalker-esque, and a feature that has to go." Within about three hours the group's membership reached 13,000. At 2 A.M. that night, it had 100,000. By midday Wednesday 280,000 had joined, and Friday it hit 700,000.

And there were about five hundred other protest groups. Their names included "THIS NEW FACEBOOK SET-UP SUCKS!!!", "Chuck Norris come save us from the Facebook news feed!," "news feed is a chump dick wuss douchbag asshole prick cheater bitch," and "Ruchi is the Devil." At least 10 percent of the site's users were actively protesting the change.

The primary objection to News Feed was that it sent too much information about you to too many people. A headline in the *Arizona Daily Wildcat* at the University of Arizona summarized: STUDENT USERS SAY NEW FACEBOOK FEED BORDERS ON STALKING. It quoted a freshman saying "You shouldn't be forced to have a Web log of your activities on your own page." And at the University of Michigan, the *Michigan Daily* quoted a junior who found it problematic on the viewer's side. "I'm really creeped out by the new Facebook," she said. "It makes me feel like a stalker." Many began referring to the service as Stalkerbook. You were stalked, and you were turned into a stalker. Who wanted that?

The company's first official reaction emerged late Tuesday night. Zuckerberg wrote a blog post with the condescending headline "Calm down. Breathe. We hear you." He took a rational line: "We're not oblivious of the Facebook groups popping up about this (by the way, Ruchi is not the devil). And we agree, stalking isn't cool; but being able to know what's going on in your friends' lives is. This is information people used to dig for on a daily basis, nicely reorganized and summarized so people can learn about the people they care about." He also noted a point that to him and his colleagues at Facebook was fundamental to News Feed: "None of your information is visible to anyone who couldn't see it before the changes."

The next day television news crews began to gather in front of Facebook's Palo Alto headquarters building. The company had to hire security guards to escort employees to and from the office. Students from several schools were calling for a massive in-person protest there. Employees were scared. "We had all these conversations," remembers Sanghvi. "'Should we shut off the News Feed?' 'Is it going to kill the company?'" There were earnest debates in Facebook's conference rooms about whether they should simply block messages about the protest groups from showing up in people's News Feeds. But Zuckerberg, in New York on a promotional trip, argued firmly with his colleagues by email and phone that this was a matter of "journalistic integrity"—to cut off debate would be contrary to the spirit of openness that led him to create the company in the first place.

But despite the hubbub, Zuckerberg and everybody else at Facebook saw one central irony about the episode: that the protest groups had grown so fast. In itself that was testimony to the News Feed's effectiveness, they believed. People were joining the groups to protest News Feed because they were learning about them in their News Feeds. As Zuckerberg explained it to me at the time, "The point of the News Feed is to surface trends going on around you. One thing it surfaced was the existence of these anti-feed groups. We really enabled these memes to grow on our system." To him it was the ultimate evidence that News Feed worked as it was intended.

However, such calm and clever logic would not quell the uprising. So Zuckerberg agreed to compromise. Cox, Sanghvi, senior engineer Adam Bosworth, and several other engineers spent a frantic forty-eight hours writing new privacy features that gave users some control over what information about them was being broadcast by the News Feed. You could now instruct the software not to publish stories about specific sorts of actions. For instance, you were able to silence it when you commented on a photo, or—and this was an important one—when you changed your relationship status.

Zuckerberg stayed up all Thursday night in his hotel room in New York writing a new blog post announcing the new privacy controls. It had a markedly different tone than his first one. "We really messed this one up," it began. "We did a bad job of explaining what the new features were and an even worse job of giving you control of them. . . . We didn't build in the proper privacy controls right away. This was a big mistake on our part, and I'm sorry for it." He also announced that in a few hours he would be participating in a live public discussion about the News Feed on a group called "Free Flow of Information on the Internet."

The "Students Against Facebook news feed" group peaked that day at 750,000 members. The demonstrations were canceled. The privacy controls tamped down the protest quickly.

The News Feed enabled very large groups to form on Facebook almost instantly. That had never been possible before. And the anti–News Feed groups were not the only ones that burgeoned that first week. Even as "Students Against Facebook news feed" was gathering steam, another one with a more juvenile tone was taking off. It was called "If this group reaches 100,000 my girlfriend will have a threesome." It reached its target in less than three days, as awareness spread via the virality of the News Feed. (The message turned out to be a hoax.) Meanwhile yet another new group was collecting tens of thousands of new members and reassuring Facebook employees that there was in fact some redeeming value to News Feed. It was called "Save Darfur."

Zuckerberg was entirely willing to tweak News Feed, but he never for a moment considered turning it off. Explains Cox: "If it didn't work, it confounded his whole theory about why people were interested in Facebook. If News Feed wasn't right, he felt we shouldn't even be doing this" ("this" being Facebook itself). But Zuckerberg in fact knew that people liked the News Feed, no matter what they were saying in the groups. He had the data to prove it. People were spending more time on Facebook, on average, than before News Feed launched. And they were doing more there—dramatically more. In August, users viewed 12 billion pages on the service. But by October, with News Feed under way, they viewed 22 billion.

The first time I ever met Zuckerberg was at lunch on Friday, September 8, the day Facebook unveiled the News Feed privacy changes. Only hours earlier he'd posted his contrite letter to users after staying up all night, and he was shortly to participate in the live question-and-answer session to help placate protesters.

He was completely unfazed. He arrived at the restaurant wearing a short-sleeved T-shirt decorated with a whimsical image of a bird on a branch. He immediately launched into a confident peroration about social networking and how Facebook fit into it; he almost disregarded the fracas he'd spent the preceding days trying to tamp down. His rhetoric was big-picture and visionary. Almost offhandedly, he shared his dispassionate analysis of why Facebook's users were so mad about News Feed. He said that he hadn't anticipated the uproar because he had thought users would realize that nothing on News Feed hadn't already been visible on Facebook in the past; it was just better organized and presented. But he now realized, he said, that this argument was only hypothetical. It was apparent that people felt that normal obstacles to intrusiveness had been improperly removed. He was starting to realize that users take time to get used to changes, no matter how inevitable or necessary they might seem to him.

News Feed was more than just a change to Facebook. It was the harbinger of an important shift in the way that information is exchanged between people. It turned "normal" ways of communicating upside down. Up until now, when you desired to get information about yourself to someone, you had to initiate a process or "send" them something, as you do when you make a phone call, send a letter or an email, or even conduct a dialogue by instant message.

But News Feed reversed this process. Instead of sending someone an alert about yourself, now you simply had to indicate something about yourself on Facebook and Facebook would push the information out to your friends who, according to Facebook's calculations of what was likely to interest them, might be interested in the activity you were recording. And for the recipients of all this information, looking at their Facebook home pages, this new form of automated communications made it possible to stay in touch with many people simultaneously with a minimum of effort. It was making a big world smaller.

In essence, what Facebook had created was a way to "subscribe" to information about a friend. Instead of waiting for a friend to send you information, now you told Facebook—merely by being friends with someone—that you wanted to hear about them. To friend them was to subscribe to their data, so that Facebook's software would pull their information to your page. The main precedent for such a subscription model was the first well-known system for feeds—RSS (Really Simple Syndication). RSS had become popular along with blogging a few years earlier. It was a way to subscribe to the output of a given blog or website. RSS feeds had become a routine way for Web denizens to receive news, commentary, and many other types of information. Applying it to behavioral information about people, however, was a radical departure for the Net and would prove to be hugely influential.

But in their anger about the News Feed students were nonetheless recognizing something important and, for many, genuinely disturbing— when people can see what you are doing, that can change how you behave. The reason the News Feed evoked something as intrusive as stalking was that each individual's behavior was now more exposed. It was as if you could see every single person you knew over your backyard fence at all times. Now they could more easily be called to account for their actions.

Facebook had acquired the power to push people toward consistency, or at least to expose their inconsistencies. Once everything you do is laid out in chronological order for your friends to see, that may allow people to recognize things about you that they never previously knew, whether for good or ill. If you smoked a joint and a friend happened to snap a photo, that photo might get posted on Facebook. If you held a party and didn't invite a friend, they were now more likely to find out about it. You were asked to declare whether you were "in a relationship" or "single." You couldn't tell one girl one thing and another girl another. Any change in your relationship status would get pushed out on the News Feed.

Another reason many Facebook users were upset with the News Feed was more unsurprising—they had accepted too many "friends." Facebook was designed as a way to communicate with people you already knew. But for many it had instead become a way to collect friends, even a competition to see who could have the most. But if your behavior was going to be broadcast to everyone on your friend list, people who had engaged in rampant friending now had little control over who saw into their private lives.

In his planning for News Feed and his response to the revolt, Zuckerberg established a pattern he would repeat in future controversies. He pushed for News Feed out of his conviction that it was the logical next step for the service. He did not give sufficient consideration in advance to how it would impact users' sense of privacy, and more importantly, how it would make them feel. Not everyone appreciated the transparency that Zuckerberg envisioned. One person's openness was another person's intrusiveness. Zuckerberg resisted criticism at first, then capitulated and turned contrite. In the end he embraced dialogue with the protesters. Facebook's iterative approach to all things prevailed. And more or less, all was well.

Despite the rocky start for News Feed, Zuckerberg considered it critical that Facebook continue expanding its reach. He still wanted to move quickly toward open registration. He wanted this not because he wanted more users so Facebook could make more money; instead he thought that Facebook was more useful as it acquired more users. At lunch on September 8 he said, "Whenever we expand the network, that makes the network stronger."

Zuckerberg also never considered mothballing the open registration plan. He and his colleagues, Chris Hughes and public relations manager Melanie Deitch, did debate among themselves at our lunch whether to open up as planned the following week or to delay open reg to allow the News Feed hullabaloo to die down.

In the end Zuckerberg delayed open registration by just two weeks, until September 26. That was partly so that additional privacy controls could be added to ensure that student users didn't feel that new, older users coming in following open registration would shadow them. He wasn't going to make exactly the same mistake twice in one month.

But there was another distraction during those same weeks that took up a lot of Zuckerberg's time—Yahoo returned. Even after the company's stock had plummeted in July and it had retreated from its billion-dollar offer, CEO Semel still badly wanted to own Facebook. He and his staff watched the explosion of the News Feed controversy and its rapid denouement as Zuckerberg deftly addressed the objections. They were impressed. Separately, Yahoo's stock had regained more than half the value it lost in July, bolstering Semel's nerve.

Now Semel reapproached Zuckerberg with the surprising news that he wanted to renew his original $1 billion purchase offer. He even suggested he might go higher. This was a new situation.

Though Zuckerberg had stayed coolheaded during the News Feed crisis, the young CEO was now unnerved. His users suddenly seemed less predictable. And the failure of the work networks continued to gnaw at him. He was losing confidence in the prospects for open registration, which would launch in mere days. And he had promised the board he would take a billion-dollar offer seriously.

Zuckerberg and Breyer had a blunt conversation. Both clearly recalled the stress of the earlier negotiations. Zuckerberg began to wonder if in fact he ought to sell the company. "I want to keep our options open," he told Breyer. "If the number of users and engagement is not growing steadily after open registration, maybe that billion or billion-one is something I'd want to do."

Open registration and the launch of the address-book importer became a make-or-break test of Facebook's long-term viability. Would it flop as work networks had? Were adults ever going to want to join Facebook?

Open registration launched on September 26. Every day for the next two weeks, a group of six pored over the latest data. The group included Zuckerberg, Breyer, Peter Thiel, COO Van Natta, "consigliere" Cohler, and co-founder Moskovitz. In the last few days of September the data was nerve-rackingly unclear, which meant that a sale might be imminent. Yahoo's lawyers were again conducting due diligence, getting ready for a deal. Sean Parker was watching closely from the sidelines, appalled. "We almost took the offer," he says. "It was the only time Mark felt he couldn't withstand the pressure from his teammates."

But Zuckerberg's confidence in Facebook's strategy was again vindicated. One colleague remembers being in the CEO's all-white private conference room during these weeks when somebody burst in and announced "Ten million! This is so great!" Reaching that many users was a major milestone in the company's growth.

After about a week it was apparent that not only were adults joining

Facebook, but once inside they were inviting friends, posting photos, and doing all the other things that active users did. They were engaged. Prior to open registration, new users were joining at a rate of about 20,000 a day, but by the second week in October the figure was 50,000. And students didn't rise up against the new adult users as some had feared. Perhaps the News Feed ruckus had worn users down. Or maybe they were so busy checking out all the stuff they were learning about on News Feed that they didn't have time to protest.

Breyer, in particular, was assuaged by the results of open registration. "Opening it up kicked in new usage," recalls Breyer. "At that point, it was pretty much game over. Our growth numbers looked good. And we just said, 'We're not ready to sell.'"

The company may have remained intact, but some of Zuckerberg's relationships did not. In the months that followed, his dealings with Breyer were strained. Van Natta had pushed so hard for a sale to Yahoo that Zuckerberg never fully trusted him again, according to one of Zuckerberg's close friends. Van Natta remained as COO for another year. Even Cohler, one of Zuckerberg's closest confidants, felt the tension. For a while Cohler was excluded from the inner circle. Says an adviser to Zuckerberg, "Mark is all about loyalty to the company, and if you want to sell the company you're not friends to Mark Zuckerberg. Mark remembers everybody who was in favor of the Yahoo deal."

But in the wake of that tumultuous September of 2006, Zuckerberg's stature as a leader soared at Facebook. Many employees even began to view him with a touch of awe. Everyone knew he had been resolute about both News Feed and open registration. Says one senior executive, speaking of Zuckerberg's response to the News Feed protests, "It was a moment of greatness for Mark. It cemented him as the person who would run this company forever. He looked at his conscience and came up with a great compromise so people could better control the information being shared. That completely silenced everyone, and within a few days the whole thing had blown over."

And while many of the company's 130 employees wondered if it made sense to turn down Yahoo—after all, many would have become multimillionaires if Zuckerberg had agreed to sell—the company's forward progress now started to acquire an air of inevitability. Board member Breyer began to allow himself to envision a much grander Facebook that spanned the entire Internet, a vision he'd resisted in the past. Naomi Gleit, a product manager who had been opposed to News Feed, voices the feelings of others: "He was just two steps ahead of everybody else," she says. "He had pushed the company, and gotten lots of negative feedback. But he had been right."

Zuckerberg himself remembers the anxiety of the Yahoo talks. "It was one of the most stressful times," he says, in an uncharacteristic acknowledgment of his own anxieties. He worried how employees would react when he and the board decided not to sell. "I was really lucky because a lot of times when a company goes through a hard decision like that it can be years until it's clear that you made the right decision. Whereas in this case it was pretty clear very quickly."

At one staff meeting during those chaotic weeks, when Facebook's ability to maintain its momentum seemed so precarious, twenty-two-year-old Mark Zuckerberg showed a candor that both surprised many of his colleagues and endeared them to him. "It may not make you comfortable to hear me saying this," he said, "but I'm sort of learning on the job here."

For its holiday party that December, the entire company, now about 150 people, took buses to the Great America Theme Park in nearby Santa Clara. From the minute people got on the buses they started drinking. By the time they arrived at the park many were already drunk. Facebook's employees celebrated a successful year on the park's thrill rides that spun, dropped, twisted and inverted them. On the way home an employee threw up in an air vent of one of the buses. The company had to pay several thousand dollars to repair the damage. It was, in a way, Facebook's last gasp of amateurism. The company had 12 million active users. It had passed the point where it could be run like a dorm-room project.

10

Privacy   "You have one identity."

How much of ourselves should we show the world? It's an important question Facebook forces us to confront. Do I want you to know that I am a longtime *Fortune* magazine journalist who covers technology and is now writing a book about Facebook? Or should I tell you I am a fifty-seven-year-old husband of an artist, father of a teenage girl, sometime poet, and former union activist? Up to now, depending on the social context, I would most likely have presented one or the other of these identities to you. On my single Facebook profile, pretty much all is revealed.

That is no accident. Zuckerberg designed Facebook that way. "You have one identity," he says emphatically three times in a single minute during a 2009 interview. He recalls that in Facebook's early days some argued the service ought to offer adult users both a work profile and a "fun social profile." Zuckerberg was always opposed to that. "The days of you having a different image for your work friends or co-workers and for the other people you know are probably coming to an end pretty quickly," he says.

He makes several arguments. "Having two identities for yourself is an example of a lack of integrity," Zuckerberg says moralistically. But he also makes a case he sees as pragmatic—that "the level of transparency the world has now won't support having two identities for a person." In other words, even if you want to segregate your personal from your professional information you won't be able to, as information about you proliferates on the Internet and elsewhere. He would say the same about any images one individual seeks to project—for example, a teenager who acts docile at home but is a drug-using reprobate with his friends.

Zuckerberg, along with a key group of his colleagues, also believes that by openly acknowledging who we are and behaving consistently among all our friends, we will help create a healthier society. In a more "open and transparent" world, people will be held to the consequences of their actions and be more likely to behave responsibly. "To get people to this point where there's more openness—that's a big challenge," says Zuckerberg. "But I think we'll do it. I just think it will take time. The concept that the world will be better if you share more is something that's pretty foreign to a lot of people and it runs into all these privacy concerns."

Most people would find these views discomfiting, and Zuckerberg spends little time dwelling on the obvious downside of his vision. The path to more openness is already strewn with victims whose privacy was unwillingly removed. As one expert in privacy law recently asked, "How many openly gay friends must you have on a social network before you're outed by implication?" The problems with privacy on Facebook typically arise when the comfortable compartments into which people have segregated various aspects of their lives start to intersect. You may attempt to project one identity for yourself on your Facebook profile, but your friends, through their comments and other actions, may contradict you.

Facebook is founded on a radical social premise—that an inevitable enveloping transparency will overtake modern life. But through strength of conviction, consistency, and strategic flexibility, Zuckerberg has been able to keep Facebook true to this premise despite the pressures that have come as it grows toward 500 million users. To understand Facebook's history you must understand Zuckerberg's views about what at Facebook they call "radical transparency." The company's most painful moments have come because it took actions—like the launch of News Feed—that suddenly exposed users' information in unexpected ways.

With its mammoth scale, Facebook's very success has rendered the premise less alarming. For better or worse, Facebook is causing a mass resetting of the boundaries of personal intimacy. A large number of Facebook's users, especially younger ones, revel in the fullness of disclosure. Many users willingly fill out extensive details about their career, relationships, interests, and personal history. If you are friends with someone on Facebook, you may learn more about them than you learned in ten years of offline friendship. Zuckerberg considers himself a strong partisan for privacy rights and is proud that Facebook has from the beginning offered users so many controls to determine who sees their information. But he also strongly believes that people are rapidly losing their interest in sequestering their data. So to keep the service in line with what he sees as changing mores, he continues to pust Facebook's design toward more exposure of information, even as most privacy controls remain in place. This contradiction helps explain the series of privacy-related controversies that have dogged the company throughout its history—around the News Feed in 2006, Beacon in 2007, the terms of service in early 2009, and the "everyone" privacy setting in late 2009. In each case the company pushed its users a bit too hard to expose their data and subsequently had to retreat.

But despite Zuckerberg's opinion there remain many ways in which social conventions and personal behavior have not yet caught up to Facebook's uncompromising environment. As it becomes harder to orchestrate how others view us, does that make us more consistent, or just more exposed? Longtime Facebook Chief Privacy Officer Chris Kelly echoes his boss: "We've been able to build what we think is a safer, more trusted version of the Internet by holding people to the consequences of their actions and requiring them to use their real identity." Outside experts take a different view. "At every turn, it seems Facebook makes it more difficult than necessary to protect user privacy," wrote Marc Rotenberg, executive director of the Electronic Privacy Information Center (EPIC) and a respected Internet watchdog, in a mid-2008 op-ed essay. Rotenberg believes that users are not given sufficiently simple controls for their information, and that Facebook for all its belief in transparency is not very transparent about what it does with our information.

The amount of data about us that resides on Facebook also raises public policy questions about privacy. Should this company—or any one company—control and aggregate so much inside its own infrastructure? Should that be a job for government? People want to be in command of their digital identity. Even if Facebook makes promises about how it will treat our data, how can we be certain it will be used as we say it should, not only now but in the future? Facebook makes the personal data provided by users available to advertisers, in aggregated form, for its own commercial gain. It and its business partners learn a lot about us, but in general we know far less about it and exactly how the company is using our data.

Privacy activist Rotenberg certainly thinks so. "Who will control our digital identity over time?" he asks. "We still want control. We don't want Facebook to control it." Facebook will certainly face repeated backlash both from users and government regulators as its privacy policy evolves.

The older you are, the more likely you are to find Facebook's exposure of personal information intrusive and excessive. Many adult users of Facebook have trouble accepting the idea that a single profile should conflate their personal and professional lives. Some of them therefore use it exclusively for genuinely personal information and try to avoid accepting friends from work. Others keep personal stuff to a minimum and connect indiscriminately with work colleagues and contacts, including those they don't know well, aiming to turn Facebook into a networking bonanza. My Facebook friend Robert Wright, fifty-two, a respected nonfiction author who recently published *The Evolution of God,* only went on Facebook reluctantly, to help promote his writing. "Facebook requires an amount of disinhibition that is not natural to me. I'm too self-conscious to use modern technology effectively," he says.

Even some of Zuckerberg's associates disagree with him. "Mark doesn't believe that social and professional lives are distinct," says Reid Hoffman, the early Facebook investor and creator of the business-only LinkedIn social network, which discourages inclusion of personal information. "That's a classic college student view. One of the things you learn as you get older is that you have these different contexts." Longtime Facebook programmer Charlie Cheever (now departed from the company) is another skeptic: "I feel Mark doesn't believe in privacy that much, or at least believes in privacy as a stepping-stone. Maybe he's right, maybe he's wrong." By "stepping-stone," Cheever means Zuckerberg sees privacy as something Facebook should offer people until they get over their need for it.

But some theorists of business applaud Zuckerberg's approach. John Hagel, fifty-nine, a top researcher and consultant at Deloitte Consulting and author of several bestselling books about the Internet and business, believes presenting what he calls "a holistic version of ourselves" is inevitable and probably beneficial. The reason, he says, is the accelerating pace of change in business and society. "If we don't keep acquiring new knowledge by participating in broader networks of relationships, we'll be out of work," he explains. "But sustained relationships must be based on trust, and that's harder if you're only showing a part of yourself."

It's not that Zuckerberg believes in total disclosure. He wouldn't reveal confidential goings-on at Facebook on his own profile. Hagel too has his limits. "If I'm going to criticize my daughters I won't do it on Facebook," he says. "On the other hand, it's valuable for people to know I have two daughters because it creates more sense of who I am as a person."

Some people thrive on the unbridled self-disclosure. Jeff Pulver, a New York tech entrepreneur and consummate networker both on and offline, does much of his business on Facebook and Twitter, using them to send messages and arrange meetings. But he also is his real self in such interactions, he insists. "I call it life 3.0," he says, "living more and more of your life online and connecting in real ways. People who have their shields up and don't make themselves vulnerable won't ever understand why there's all this excitement about Facebook and Twitter and social media."

In 2007, London-based technology expert Leisa Reichelt coined the phrase "ambient intimacy" on her blog to describe the dynamics of Facebook and other new services that enable individuals to freely talk about themselves to groups of friends or followers. She defined it as "being able to keep in touch with people with a level of regularity and intimacy that you wouldn't usually have access to, because time and space conspire to make it impossible." The phrase struck a nerve globally with students of social networks. A widely discussed 2008 article in the *New York Times Magazine* by Clive Thompson detailed his own experience with Facebook and Twitter. It explored the social implications of ambient intimacy and was an argument for its virtues. "The new awareness . . . brings back the dynamics of small-town life, where everybody knows your business," Thompson wrote, approvingly.

The reality is that nothing on Facebook is really confidential. The company's own privacy policy is blunt on this score. Any of your personal data "may become publicly available," it reads. "We cannot and do not guarantee that User Content you post on the Site will not be viewed by unauthorized persons." To be fair, this language is intended primarily to inoculate Facebook against potential lawsuits. The company certainly tries hard to give you protections for what is meant to be confidential. But many people do not understand or take advantage of Facebook's often-complicated controls for their own information. That frequently leads to misunderstandings and embarrassment.

Once people expose their real behavior on Facebook, when they do something rash or stupid it is more likely to become "publicly available." A young U.S. employee of Anglo-Irish Bank asked his boss for Friday off to attend to an unexpected family matter. Then someone posted a photo on Facebook of him at a party that same evening holding a wand and wearing a tutu. Everyone in the office—including his boss—discovered the lie. A political candidate in Vancouver, Canada, withdrew from his race after a newspaper published a Facebook photo showing two people happily pulling on his underwear. Notoriously, Barack Obama's speechwriter Jon Favreau was publicly embarrassed when a blog published a photo that showed him at a party with his hands on the breast of a life-size cardboard cutout of Hillary Clinton. It had been posted on Facebook by one of his friends. And Facebook disclosure can do more than merely embarrass you. A 2009 poll of U.S. employers found that 35 percent of companies had rejected applicants because of information they found on social networks. The number one reason people weren't hired: posting "provocative or inappropriate photographs or information." Colleges too are increasingly searching Facebook and MySpace as they make admissions decisions.

Perhaps the Favreau incident was on President Obama's mind when he spoke to a group of high school students in Virginia in September 2009. "I want everybody here to be careful about what you post on Facebook," he said, "because in the YouTube age, whatever you do will be pulled up later somewhere in your life. And when you're young, you make mistakes and you do some stupid stuff." Facebook membership is becoming common among younger and younger children—it is now commonly used by many eleven-year-olds and those even younger, despite Facebook rules that users must be thirteen.

You don't have to be young to make mistakes there, however. Numerous Facebook incidents have exposed unseemly behavior by people in positions of responsibility. A guard at a Leicester, England, prison was fired after colleagues noticed he was friending prisoners. A Philadelphia court officer was suspended and reassigned after a juror in his courtroom reported he had asked her to be his Facebook friend. Jurors also have erred. Several verdicts in various parts of the United States have been challenged by convicted defendants after they learned that supposedly silenced jurors had posted remarks on Facebook while the trial was under way.

Even people whose very job is to keep secrets are flummoxed by Facebook's inducement to transparency. After the United Kingdom announced in mid-2009 that Sir John Sawers would become the next head of its spy agency, the Secret Intelligence Service (formerly called MI6), the *Daily Mail* newspaper discovered a publicly accessible trove of family photos that had been posted by his wife on Facebook. They included images of holidays, family friends, and details that could reveal where Sawers lived and how he spent his time.

Facebook transparency can jar intimate relationships. Many still haven't gotten used to seeing and knowing so much about their significant others. If your boyfriend shows up in photos with another girl, it may mean nothing, but who knows? Worse is when someone learns they are no longer a couple at all—by seeing a change in a Facebook profile. The outcome can even be tragic: a British man allegedly killed his wife, from whom he had recently separated, after he saw her relationship status on Facebook change from "married" to "single."

Photos in particular can reveal, as they did for Sir Sawers, who you spend time with, what you do with them, and where you go. High school and college students essentially conduct their lives in the open on Facebook. They conduct one-to-one dialogues with their friends on their Facebook "wall" despite the fact that anyone else with access to that profile can see it. This information is generally visible to anyone in their school network.

A few dissenters in the young generation find the obsession with Facebook self-presentation unhealthy. Shaun Dolan, a twenty-five-year-old New York assistant in a media firm, has made a deliberate decision to stay off the service. "My generation is unbearably narcissistic," he said in an email to me. "When I go out with my friends, there is always a camera present, for the singular goal of posting pictures on Facebook. It's as if night didn't happen unless there's proof of it on Facebook. People painstakingly monitor their own Facebook page to see what pictures they get tagged in, or what picture would best represent them to their friends."

Some call such behavior exhibitionism, or, as my longtime *Fortune* colleague Brent Schlender puts it, a search for "digital fame." On Facebook we follow the minutiae of our friends' lives the same way millions follow Britney Spears in *People* magazine. Andy Warhol famously said that "everybody will be famous for fifteen minutes," but on Facebook what's limited is not how long you are famous but how widely. It may be only among a circle of friends or school-mates. The Internet theorist David Weinberger now posits that "on the Web, everybody is famous to 15 people."

Many young people don't seem to know when extreme self-exposure becomes reckless. A twenty-year-old employee of Petland Discounts in Akron, Ohio, posted a photo of herself on Facebook holding two rabbits she had just drowned. Animal rights activists were outraged and she was shortly arrested and charged with cruelty to animals. Teenagers routinely post photos showing themselves and others using drugs or drinking when they are not of legal age. At Amherst Regional High School in Amherst, Massachusetts, a student gathered up pictures that showed popular kids drinking and possibly using marijuana, then sent them en masse to the school principal and others in the community. At another high school, the principal went onto Facebook and suspended all the athletes he saw in photos of a party who were holding bottles of beer. (Those with red plastic cups were spared.)

Facebook interactions with teenagers are almost universally fraught for adults, because the two generations have such fundamentally different attitudes about what is proper personal disclosure. One San Francisco executive was friended by her partner's teenage son. When he took a summer trip to Europe he headed to Amsterdam and excitedly told friends on Facebook all about his pot smoking. My friend was torn—should she tell her partner, or would that be betraying the trust given her by the teenager? A sixty-year-old in Virginia saw her nephew swearing furiously on his Facebook page, but knew that his extremely strict school could expel him for that. She confronted him about it herself rather than telling his parents.

Since most teenagers still won't friend their parents, some families have instituted a rule that as a condition of having a computer and using Facebook the parents get access to their child's profile. They are frequently distressed by what they find there.

How much Facebook should encourage users to reveal has been the subject of debate throughout the company's history. "Our mission since day one has been to make society more open," says marketer Dave Morin, a member of Zuckerberg's inner circle. "That's what it's all about, right? We help people be more open across more contexts. I think they have to worry less all the time about being who they actually are." But Facebook COO Sheryl Sandberg, thirty-nine, looks at it slightly differently. "Mark really does believe very much in transparency and the vision of an open society and open world, and so he wants to push people that way," she says. "I think he also understands that the way to get there is to give people granular control and comfort. He hopes you'll get more open, and he's kind of happy to help you get there. So for him, it's more of a means to an end. For me, I'm not as sure." Sandberg, fourteen years Zuckerberg's senior, thinks it's fine if someone doesn't want to make his or her life transparent.

Facebook does have a unique ability to help users control where information about themselves flows. But it only works because of Facebook's rigid requirement that people use their real names. If you weren't confident people on Facebook were who they said they were, you would not be able to selectively permit them to access your data by friending them. You can restrict or amplify the extent of their view into your information, as well as adjust how much information you see about them, by putting them into groups called Friend Lists. These groups—for work, family, college friends, or whomever—enable you to send information to one group and not to others. However, only about 25 percent of users actively use these controls, according to Facebook's chief privacy officer, Chris Kelly. Many consider them maddeningly difficult to use.

Facebook at least potentially already has more ways for users to control their data than just about any other site on the Net. Longtime top company architect Adam D'Angelo says Facebook represents a "new model for information" because of these controls. "Every piece of information on Facebook is protected by restrictions that say who can see it," he says. "Certain sets of people can see certain pieces of information." D'Angelo is right to note that such "granular" controls are found almost nowhere else on the Net, partly because only Facebook has so much information about who is doing the looking.

In late 2009 Facebook renovated its privacy controls and made a major effort to explain to users how to put friends into groups and assign various levels of disclosure to information. However, in the course of requiring users to adjust their settings, the company set the default setting on new controls to "everyone." Many users who were not paying attention found their information more exposed rather than less, despite this supposed "improvement" in privacy. The counterreaction was strong. A group of privacy organizations led by Marc Rotenberg and EPIC filed a formal complaint with the U.S. Federal Trade Commission, asking for an investigation and penalties for Facebook. The complainants included important groups like the American Library Association and the Consumer Federation of America. Before the change Facebook executives had spoken enthusiastically of it, saying it was likely to reassure users about their data. Ironically, EPIC's suit asserted quite the contrary: "Facebook's changes to users' privacy settings disclose personal information to the public that was previously restricted . . . These changes violate user expectations, diminish user privacy, and contradict Facebook's own representations." The company had still not learned how to anticipate and accommodate the concerns and attitudes of its users regarding privacy. They apparently were not yet ready for too much transparency.

Zuckerberg more or less lucked into giving Facebook's users the control they do have. In the beginning it became apparent that users at Harvard shared so much about themselves because they knew that only other Harvard students—members of Facebook's Harvard network—could see it. So as Facebook evolved, the concept of networks grew with it. All users were initially put into a network by default—for a university, a high school, a workplace, or a geography. For years I was in the Time Inc. network and also the New York one. You can see information about other people in your networks, and they can see yours unless you adjust your privacy preferences to prevent it. (I do, for both networks.) But nobody outside the network can see your information unless you explicitly permit them to. Now, in a key change, regional networks are being eliminated. That will dramatically reduce the number of people who can see most users' data if they haven't "friended" them.

For all the privacy challenges on Facebook, most people seem comfortable with how it works. In a September 2009 survey it was found to be the tenth-most-trusted company of any type in the United States in a survey of 6,500 consumers by research firm Ponemon Institute and TRUSTe, which verifies Internet sites. Facebook ranked ahead of Apple, Google, and Microsoft.

But the influence of Zuckerberg's more extreme convictions remains apparent as you walk Facebook's halls. Some there talk about a concept they call either "ultimate transparency" or "radical transparency." Since the world is likely to become more and more open anyway, people might as well get used to it, the argument goes. Everything is going to be seen.

The place where your information is most obviously transparent is Facebook's photos application. That's where it is hardest to limit the disclosure of information about yourself. You have no control over whether someone posts a photo of you there. You do have the right to delete the "tag" on a photo that identifies you and causes that information to be disseminated to your friend list. However, generally by the time you delete one, news of the tag has already been distributed in Facebook's News Feed. (Any user can also adjust Facebook's privacy settings so they cannot be tagged at all.) Photos are visible by default. Everyone on the entire service can see them unless you deliberately adjust your privacy controls, and most users don't.

Many users over the years have wanted Facebook to remove objectionable photos of them taken by others. However, the company follows a firm policy that while the tag is in your control, the photo is not. It belongs to the photographer. Facebook has also, wrongly in my view, resisted letting users approve tags of themselves before they are affixed to a photograph and distributed to friends.

Proponents of radical transparency argue that while Facebook may make it easier for people to see photos of you, there are many other sites on the Internet where a photographer could also post those photos. So Facebook is not facilitating anything that might not happen anyway.

"Mark's view is that Facebook had better not resist the trends of the world or else it'll become obsolete," says the soft-spoken but passionate Adam D'Angelo, who shares this view and with whom Zuckerberg has discussed such issues since they were at Exeter in 2001. "Information is moving faster," he continues. "That's just how the world is going to work in the future as a consequence of technology regardless of what Facebook does." Even Sheryl Sandberg takes evident pride when she says, "You can't be on Facebook without being your authentic self."

Members of Facebook's radical transparency camp, Zuckerberg included, believe more visibility makes us better people. Some claim, for example, that because of Facebook, young people today have a harder time cheating on their boyfriends or girlfriends. They also say that more transparency should make for a more tolerant society in which people eventually accept that everybody sometimes does bad or embarrassing things. The assumption that transparency is inevitable was reflected in the launch of the News Feed in September 2006. It treated all your behavior identically—in effect telescoping all your identities, from whatever context, into the same stream of information.

Those who speak their minds and show themselves on Facebook sometimes do see themselves as waging small battles for openness and transparency. Some of the controversies that result shine a spotlight on closed-mindedness by adults. Kimberley Swann, a sixteen-year-old in Essex, England, got a new job as a marketing firm office administrator. She added some co-workers as Facebook friends. After a few weeks she wrote on Facebook that her job was boring. Someone showed her boss, who promptly fired her. "I didn't even put the company's name," said Swann in an interview with the *Daily Telegraph*. "They were just being nosy, going through everything." Added a union official quoted by the BBC about the widely covered incident, "Most employers wouldn't dream of following their staff down to the pub to see if they were sounding off about work to their friends."

A few high school students have gone to court to defend their right to speak freely on Facebook. Katherine Evans, a student at Pembroke Pines Charter High School in Florida, created a Facebook group complaining that her Advanced Placement English teacher was "the worst teacher I've ever met." The principal learned of the group and suspended her for three days. She then sued the principal in federal court, arguing he had violated her First Amendment right to freedom of speech.

Some young people—inadvertently echoing Zuckerberg—say it's not a problem to have libertine images of themselves on Facebook because as they get older, standards about such indiscretions will have relaxed. While they are clearly gambling with their own reputations, the inarguable wholesale movement toward self-disclosure on Facebook and even in broader society gives this view some credence. President

Barack Obama openly admitted in his autobiography to having snorted cocaine. Almost nobody cared.

It's understandable that people would want to share information about themselves unreservedly and still feel protected from inadvertent disclosures that might embarrass them. But the reason they can't is embedded in the very reason people use Facebook. James Grimmelmann, an associate professor at the New York Law School, explains this dilemma in a 2009 article titled "Saving Facebook": "[Facebook] has severe privacy problems and an admirably comprehensive privacy-protection architecture. . . . Most of Facebook's privacy problems are . . . natural consequences of the ways that people enthusiastically use Facebook." He also writes, "There's a deep, probably irreconcilable tension between the desire for reliable control over one's information and the desire for unplanned social interaction."

One of Grimmelmann's central points is that the violations of privacy that occur on Facebook are frequently the result of the behavior not of the company but of people a user has accepted as a friend. To prevent photos from being taken and posted on Facebook, some college parties now ban cell phones and cameras. Some parties even have what kids call "shot rooms," which are totally dark so nobody can take any pictures of drinking or drug use. Athletes and other students concerned about their image have also learned to quickly troll Facebook after incriminating parties, seeking tagged photos of themselves that they, of course, detag. But the only way those photos can be uploaded and the tags affixed in the first place is if it's done by a user who is your "friend." Grimmelmann calls this sort of thing "peer-to-peer privacy violations."

Because we use our real names on Facebook, we can be held responsible for what we say. Many on the Internet take shelter behind pseudonyms when they say something obnoxious, rude, or hateful, but that's harder here. In Harrison, New York, a police detective was demoted and forced to retire early in 2009 after writing on Facebook that the election of President Obama meant "the rose garden will be turned into the watermelon garden."

Facebook's culture of accurate identity is not foolproof. Many create fake profiles for fun. At any time there are scores of profiles under the name Haywood Jablomie, for instance. But such fakesters are usually obvious. We are validated in our identity by the friends we have on Facebook, and Haywood usually has few or none. Other fake profiles are harder to detect. The Symantec security software firm conducted an experiment in 2008 in which it created one for an attractive young woman who supposedly attended a high school in Silicon Valley. Within hours a number of boys at that school had sent her friend requests, presumably because they wanted to date her. Sad incidents have also emerged in which, for example, men have posed as attractive women in order to get boys to send them photos of themselves nude or having sex.

Celebrities also break the Facebook model. Microsoft chairman Bill Gates shut down his personal profile on Facebook in early 2008 for two reasons. He was getting more friend requests per day—thousands—than even his staff could manage. But there were also five other "Bill Gates" profiles pretending to be him, each with numerous "friends."

People with unusual names have a different problem. Facebook often blocks their efforts to establish profiles in the first place. An Australian woman named Elmo Keep, twenty-seven, was ejected from Facebook until she sent the company copies of her passport and driver's license. V Addeman, fifty-two, of Costa Mesa, California, tried to join Facebook but was rejected by its software. He had a lengthy argument with Facebook customer service to convince them that his legal first name is a single letter. Others who have had difficulties include Japanese author Hiroko Yoda, Rowena Gay of New Zealand, and people whose names included Beaver, Jelly, Beer, and Duck. Even Caterina Fake, the well-known co-founder of Internet photo site Flickr, couldn't initially join Facebook. (Facebook's procedures for remedying such misunderstandings were grossly inadequate until late 2009, when a more formal appeals policy was inaugurated.)

The vast majority of users identify themselves accurately. That gives Facebook some unique and practical capabilities. A man in Cardiff, Wales, located a half brother he hadn't seen in thirty-five years merely by searching for him by name in Facebook. Such family reconnections are becoming almost routine in the age of Facebook.

Many people no longer exchange email addresses and cell-phone numbers; they just look each other up on Facebook. This simple directory capability is one of its most undeniable virtues. People who are not on Facebook are increasingly seen, among some groups, as unreachable by friends and acquaintances.

Is there a risk that once a fact about us has been revealed on Facebook we may never be able to escape it? Will we always be remembered as the drunken guy wearing the funny hat in some "friend's" photo gallery? Will it become harder to evolve as people because opinions about us have already hardened? From time immemorial, people have moved to new towns and started over to escape some fact or impression about themselves that made them uncomfortable. Will that no longer be possible?

It makes sense to be cautious about how much of your data you expose on Facebook. I myself abide by the simple "front page" rule. I'm relatively comfortable exposing a large portion of my whole self to scrutiny, so I put up extensive and accurate information on my profile and actively participate in dialogue. But I try never to include anything I would be devastated to find published on the front page of my local newspaper.

Zuckerberg has acquired a surprising ally in his campaign for openness and transparency—Ben Parr, the student at Northwestern University who launched "Students Against Facebook news feed," the protest group that catalyzed the big privacy crisis. In September 2008, Parr, now a technology writer, effectively recanted. "Here's the major change in the last two years," he wrote in an article. "We are more comfortable sharing our lives and thoughts instantly to thousands of people, close friends and strangers alike. The development of new technology and the rocking of the boat by Zuckerberg has led to this change. . . . News Feed truly launched a revolution that requires us to stand back to appreciate. Privacy has not disappeared, but become even easier to control—what I want to share, I can share with everyone. What I want to keep private stays in my head."

In December 2003, the residents of two adjoining rooms in Harvard's Kirkland House posed for this photo. From left: Joey Siesholtz, Mark Zuckerberg, Billy Olson, Dustin Moskovitz, Joe Green, and Arie Hasit. (Chris Hughes is missing.) (Photo courtesy Dustin Moskovitz)

Zuckerberg & Co. launched Thefacebook in February 2004 just for Harvard students. The opening screen even then hinted at wider ambitions. (Image courtesy Facebook)

Zuckerberg in his dorm room in Harvard's Kirkland House, gazing at Thefacebook on the computer where not long before he had created it. (Photo Lowell K. Chow/*The Harvard Crimson*)

The three roommates and partners at Harvard, spring 2004. From left, Moskovitz, Chris Hughes, and Zuckerberg.
(Photo Ravi Ramchandani/*The Harvard Crimson*)

Eduardo Saverin, a year ahead of Zuckerberg at Harvard, was a math whiz with business experience who helped fund Thefacebook's launch, but later fell out with the other founders. (Photo Ravi Ramchandani/*The Harvard Crimson*)

At Thefacebook's 2004 summer rental in Palo Alto, the dining room table was the nexus. From left: Andrew McCollum (a friend who wasn't working much on Thefacebook), Zuckerberg, and intern Stephen Dawson-Haggerty. (Photo Aaron Sittig)

At a 2005 company party: Ezra Callahan, Facebook's sixth employee (depending on how you count), with Zuckerberg and, on right, Kevin Colleran, ace advertising salesperson. (Photo courtesy Kevin Colleran)

In the first real office, on Emerson Street in Palo Alto. Underneath David Choe's looming murals are, from left: Zuckerberg, Moskovitz, and Sean Parker, then Thefacebook's president. (Photo Jim Wilson/*The New York Times*/Redux)

When Facebook, now without "the" in its name, moved around the corner to a fancier Palo Alto office in September 2005, Zuckerberg wanted to retain the funky feel. He again hired graffiti artist David Choe. (Photo Kevin Colleran)

Matt Cohler, left, arrived at Thefacebook in early 2005 and brought a much-needed professionalism, serving as Zuckerberg's "consigliere." Laughing at right in the background is designer Aaron Sittig, another early employee who was influential (and still is). (Photo Ben Blumenfeld)

Zuckerberg and Parker hired Kevin Colleran, left, without meeting him in person. This picture of him with rapper 50 Cent was on his Facebook profile page. Zuckerberg thought he'd hired 50 Cent. (Photo courtesy Kevin Colleran)

One of many overcrowded offices, on University Ave. in Palo Alto in 2007. From the beginning Facebook has had trouble keeping up with its own growth. (Photo Kevin Colleran)

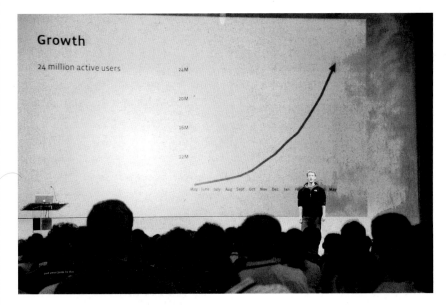

In May 2007, Facebook announced its biggest change yet—becoming a platform for applications created by others. Zuckerberg spoke at the San Francisco event called f8 (programmers wrote apps afterward for eight hours). Following f8, growth came even faster. (Photo Andrew Bosworth)

On February 4, 2008, a sea of people filled Bogota, as well as every other city in Colombia, in protest against the FARC terrorist group. It all started on Facebook, in Oscar Morales's upstairs bedroom in Barranquilla. (Photo by Alberto Acero/courtesy *Periodico El Tiempo*)

Sheryl Sandberg left Google in early 2008 to help turn Facebook into a real business. She and Zuckerberg established an effective working relationship. Revenues have grown to near $1 billion annually. (Photo courtesy Facebook)

11

The Platform "Together, we're starting a movement!"

Mark Zuckerberg has had a particular obsession since Facebook's early days. On the night that his early collaborator Sean Parker first met Zuckerberg at that trendy Tribeca Chinese restaurant in May 2004, the two got into a curious argument. Zuckerberg, in Parker's opinion, kept derailing the discussion by repeatedly talking about how he wanted to turn Thefacebook into a platform. What he meant was that he wanted his nascent service to be a place where others could deploy software, much as Microsoft's Windows or the Apple Macintosh were platforms for applications created by others. Parker argued that it was way too early to think about anything like that.

Kevin Efrusy of Accel Partners has a similar recollection. At one of his very first meetings with Zuckerberg after Accel invested in the company in late spring 2005, the young CEO asked for a favor. "Kevin, I need to find someone to help me think through my platform strategy."

"Huh? Yeah, well maybe someday we can be a platform," Efrusy replied, haltingly. "But we're just a company with six people . . . I mean, I guess I know a guy over at BEA [a business software company] who has done some interesting platform work . . ."

Zuckerberg cut him off. "BEA? I was thinking more like Bill Gates. Can you help me talk to Bill Gates?"

"Ummm . . . I don't know. Maybe Jim Breyer can help with that . . ."

A week passed. Efrusy was again at Zuckerberg's office. "Okay," said Zuckerberg. "So I talked to him."

"Talked to who?"

"Bill Gates!"

Even in these early days, Zuckerberg was trying to imagine how his little service could be more than just an Internet destination where people went to communicate with each other.

Every great technology company goes through one or two key transitional moments when its creators discover they have created something different—and bigger—than they initially realized. Early on it dawned on Bill Gates—then making bespoke software for little PC hardware companies with partner Paul Allen—that software should be its own industry. He later had a second epochal realization: that entire computers could be built around an operating system. Microsoft subsequently became the most profitable company in history. One night it hit Yahoo founders Jerry Yang and Jeff Filo that they didn't have only a map to the Internet. Their service could also be an unprecedented way to collect detailed market research about Net users. Yahoo became the first big ad-supported Net media company. Google's turnabout came when founders Sergei Brin and Larry Page discovered they could direct user searches not only toward websites but also toward a separate database of advertising. Thus was born the most powerful business model of the Internet era so far.

Zuckerberg's first eureka moment was when he and Moskovitz realized their service could go beyond college. But another struck while watching the stunning success of the photos application. It became apparent something special was happening. "Our photo site lacks features anyone else would build," Zuckerberg told me in early May 2007. "We don't store high-resolution photos. The printing function is downright bad. And until recently you couldn't even change the order of photos in an album. Yet somehow, this application became the most trafficked photo site on the Internet, by far." And something similar was going on with the application Facebook engineers had quickly thrown together to allow users to invite friends to events. It was garnering more usage than Evite.com, which had been for years the leading website for invitations.

"So why were photos and events so good?" he asked. "It was because despite all their shortcomings they had one thing no one else had. And that was integration with the social graph." This was Facebook's own conceptual breakthrough, and Zuckerberg was proud of the term he used to describe it. "We did some thinking and we decided that the core value of Facebook is in the set of friend connections," he continued. "We call that the social graph, in the mathematical sense of a series of nodes and connections. The nodes are the individuals and the connections are the friendships." Then his enthusiasm veered, it seemed at the time, toward overstatement: "We have the most powerful distribution mechanism that's been created in a generation." Zuckerberg immodestly explained that this same power could be applied to any sort of application—not just photos or events. His certitude was jarring.

By "distribution" he meant that by connecting with your friends on Facebook you had assembled a network, this so-called social graph, and it could be employed to distribute any sort of information. If you added a photo, it told your friends. Ditto if you changed your relationship status, or announced that you were heading to Mexico for the weekend. But it could also tell your friends about any action you took using any software to which your social graph was connected. So far, though, the only applications that took advantage of this distribution capability were photos, events, and a few others created by Facebook itself.

Most software companies, were they to conclude that they had such an ability to create uniquely powerful applications, would create more of them. They might make shopping applications on top of their social graph, or games, or applications for businesses. Instead, Facebook stopped building applications at all, at least for a while. In the fall of 2006 Zuckerberg set out to realize his long-held vision of a platform for others to build applications on top of Facebook. He wanted to do for the Web what Gates did for the personal computer: create a standard software infrastructure that made it easier to build applications— this time, applications that had a social component. "We want to make Facebook into something of an operating system, so you can run full applications," he explained.

COO Owen Van Natta, whom I also talked to in May 2007, had his own way of describing this potential: "Take anything today on the

Internet and overlay a lens that is people you know and trust who have their own perspective. That's what we will enable with platform. What wouldn't potentially be more valuable when seen through that lens?"

Zuckerberg had thought about platforms almost since he first touched a keyboard. He learned to program as an adolescent by coding functions that worked on top of AOL, then the dominant online service. A community of hackers—including Zuckerberg—turned AOL into a platform whether its leaders wanted it to be one or not. Then when he was a senior at Exeter, he teamed up with Adam D'Angelo and built his software for listening to MP3s (audio files) called Synapse. Synapse became popular in part because it allowed other programmers to build companion programs, called plug-ins, that supplied additional features. Synapse was, in effect, a mini-platform. And in his earlier, abandoned obsession with his cherished Wirehog, Zuckerberg was thinking of Facebook as a platform. Wirehog was in effect, if only briefly, the first independent application to operate on top of Facebook.

Becoming a platform on which the applications of others can operate is one of the great holy grails of technology. Microsoft dominated the technology industry for almost two decades by positioning its Windows software as the monopoly operating system platform for the PC industry. Anyone who wanted to build a PC application had to use Windows. (It was Bill Gates in fact who popularized this use of the word "platform.")

Creating a platform enables a software company to become the nexus of an ecosystem of partners that are dependent on its product. And once a company is at the center of an entire ecosystem, it becomes maddeningly difficult for competitors to dislodge it. Not only did Apple succeed at this masterfully with its Macintosh operating system, but it succeeded again, first with the iPod and then with its magnificent iPhone.

By becoming a platform, Facebook also takes some of the burden off itself to excel in everything it does. Facebook will never be able to build the best application in every area its users are interested in. Companies that devote more resources to chat, for example, will continue to outpace Facebook. I recently asked my seventeen-year-old daughter, Clara, if she used Facebook's chat application, an ambitious add-on the service launched in mid-2008. No, she said, she still preferred AIM and Apple's iChat (an answer many American teenagers would give, despite their addiction to Facebook). "Facebook Chat feels like using Morse code," she explained. It doesn't have enough features and isn't easy enough to use. Zuckerberg decided that what Facebook did uniquely well was maintain your personal profile and your network of friend connections. Ultimately, almost everything else will be done by other companies.

Facebook made the first move to turn itself into a platform back in August 2006. The world barely noticed. The big news around then was the News Feed scandal. Programmer Dave Fetterman spearheaded something called the Facebook API, or application programming interface. It enabled users to log in to other sites on the Web with their Facebook username and password so the partner site could extract their data, including their list of friends. Some at Facebook—mostly older executives—objected to letting user data escape the confines of the service in this way. They said the company was giving away something valuable and getting nothing in return. But Zuckerberg pushed it through. To demonstrate the API, Facebook built its own external website application called Facebank, later renamed Moochspot, for keeping track of small debts between friends.

While thousands of developers did fiddle around with the API, not many used it, and very few Facebook users did.

The real problem with the API was that it didn't help outside application partners very much, because it didn't include that vaunted "distribution." It didn't take full advantage of the social graph. You could pull your list of friends out of Facebook but you couldn't send information you produced back inside to them. You and your friends could keep track of debts on Moochspot, yet it didn't send any information back to your profile.

But shortly Facebook triumphed with News Feed. It enabled your friends to easily learn about your Facebook activities—including which applications you installed on your profile. Only with the News Feed in place could Facebook become a successful platform. Open registration also helped lay the groundwork. Software developers would obviously be more interested in applications on Facebook if it operated at a large scale and included all sorts of people.

As soon as the News Feed brouhaha settled down, the company's priorities turned to building the platform. D'Angelo and Charlie Cheever did much of the critical programming work. Dave Morin got the job of "platform marketing"—working with potential developers. (In his previous job at Apple, already a Facebook partisan, he had sought futilely to get Facebook built into the Mac OS.) Morin and Fetterman visited companies that had successfully created platforms, including eBay, Apple, and Salesforce.com.

But despite all the external models, the team kept harking back to one internal reference point. "We used photos as the model the entire time," says Morin. "We just kept looking at it, asking, 'How do we enable every application to do what photos does?'" Each profile page included a box for photo albums. Clicking on a photo took a user to an entire page, which looked much like a website. When you uploaded a photo it updated your personal mini-feed on your profile as well as the News Feeds of relevant friends. So the team decided to allow outside developers similarly to place boxes on profile pages and to build full pages inside Facebook. Actions in any application could, of course, generate News Feed stories.

Carrying this logic even further, they arrived at the principle that Facebook should not be able to do anything with its own applications that outside developers couldn't do. It should be a level playing field, Zuckerberg explained in 2007. "We want an ecosystem which doesn't favor our own applications," he said. This policy was followed to such an extreme that features were removed from Facebook's own photos application because an outside developer would not have been able to include them.

The company extended an extraordinary degree of freedom to its new partners. Amazingly, it planned to let developers make money with their applications, but would not charge them anything at all for the right to operate inside Facebook. "People can develop on this for free," said Zuckerberg around the time the platform debuted, "and can do whatever they want. They can build a business inside of Facebook. They can run ads. They can have sponsorships. They can sell things, they can link off to another site. We are just agnostic. There are going to be companies whose only product is an application that lives within Facebook."

But did it make Facebook a better business? That was not a priority. "We don't force ourselves to answer the question how we're going to make money off this right now so long as it's strengthening our market position," he said back then. "We'll figure that out later."

So Zuckerberg saw it. But some of his colleagues, particularly the ones who sold advertising for the site, were apoplectic. Why should its application partners be allowed to compete with Facebook itself in selling ads? There were plenty of angry meetings. But for all the venting, Zuckerberg was unswayed. Activity on applications, he argued, would generate more activity in Facebook. That would create more page views, and even on application pages Facebook would reserve space to sell its own ads. Zuckerberg also advocated a sort of corporate Darwinism. He said he wanted outside apps to help keep Facebook honest by forcing it to make its own remaining applications good enough to compete successfully.

Back then I talked to Zuckerberg in his private retreat—an all-white conference room furnished with midcentury modern furniture from Design Within Reach, a few blocks down University Avenue. (He didn't decorate it himself, but he liked it.) White Eames chairs, a white Saarinen table with delicate metal legs, white curtains, white blinds, gray rug and sofa, and a big black beanbag chair. Employees called it the "interrogation room" both because Zuckerberg was known for his probing questions and because its austerity evoked a prison cell. It was Zuckerberg's twenty-third birthday when he and I met there. He was barefoot and unshaven, wearing an A&W Root Beer T-shirt with blue jeans. In a corner was an unopened box of Transformer robot toys.

Zuckerberg was drawing diagrams on the whiteboards that covered all the walls, and at one point couldn't find an eraser. So he picked up a knit hat from the floor and wiped the board with that.

In April, Zuckerberg had given a talk at a News Corp. executive summit at a resort at Pebble Beach, two hours south of Palo Alto. Rupert Murdoch had recently said a few things in public to suggest he wondered if he had bought the wrong social network. At a gala dinner Zuckerberg and Murdoch huddled together intently, while MySpace CEO Chris DeWolfe sat nervously at a nearby table. Finally Zuckerberg got up, announcing that he had to get back to take his girlfriend to a movie. "After he left, the MySpace guys rushed over to Rupert," says blogger and author Jeff Jarvis, who attended the dinner. "It was like 'Dad! Pay attention to me!'"

Now as the platform launch approached, Zuckerberg made no bones about the fact that it was intended partly to best MySpace, which remained the dominant American social network. MySpace had recently decreed that some third-party applications could not operate there, and even shut one down merely on suspicion that it had been selling advertising. "We just have such a different philosophy and view of the world," Zuckerberg explained. "We're a technology company. MySpace is a media company, and they view their job as owning and distributing content."

To succeed with the platform launch, Facebook had to start promoting itself to developers. Dave Morin and Matt Cohler crisscrossed the globe visiting start-ups and big media companies alike, seeking to convince them to make software for Facebook. A splashy launch event was planned for May 24, 2007, at a big hall in San Francisco. Facebook called the event f8, a name that subtly proclaimed it was Facebook's "fate" to become a platform. Zuckerberg even emerged from his shell to solicit advance attention from a journalist, me, whom he invited inside the company for an exclusive story as he prepared for f8. I published an article titled "Facebook's Plan to Hook Up the World" in *Fortune* magazine and online at the very moment f8 began.

Facebook hired a veteran event planner named Michael Christman to oversee the f8 logistics. On his first visit to the offices he was in a lengthy meeting, sitting by the door in a big conference room that also held a flat-screen TV and a Nintendo Wii machine. The door opened and banged into Christman's back. Two young men appeared but backed out when they realized the room was in use. A few minutes later they came in again, hoping the meeting was over, and again knocked his chair. They wanted to play the video game. When it happened a third time, Christman turned and said sternly, "Boys, if you want to play with the Wii, come in. But don't bang my chair again." At that point, Meagan Marks, a Facebook employee who was managing f8, said, "Michael, this might be a good time to introduce you to our CEO, Mark Zuckerberg."

The days leading up to f8 were a frenzy of excitement and near panic. Employees were fueled by a sense that they were making history. The graffiti-scrawled halls were abuzz with grand proclamations. "We're gonna change the Internet!" "We're gonna make the Internet social!" "We're gonna finally put people on the Internet!" "We're creating a real economy on the Web!" Apple veteran Dave Morin remembers driving home one night at 4 A.M. after a particularly intense planning session, thinking, "This is what it must have been like building the first Macintosh." To prepare for the new Facebook, Morin was reading *Democracy in America* by Alexis de Tocqueville, the classic nineteenth-century observation of the U.S. political and economic system, as well as Adam Smith's *The Wealth of Nations*. Modesty of ambition has never characterized successful leaders at Facebook.

It had been a marathon of programming. Adam D'Angelo and his team building the platform worked seven days a week for more than three months. The night before f8 they were almost—but not quite—ready. A core group crowded into a room at San Francisco's W Hotel running through final fixes. Most hadn't slept for days. But a key piece of the platform software still didn't work properly.

Some of the programmers took an alertness drug called Provisual so they could stay up yet another night. They were semidelirious. They joked they should mix Provisual with cocaine and call it Blow-visual.

Luckily the quality of their coding was higher than that of their humor. But they made it through the night. Just hours before f8 was scheduled to begin, they flipped the switch. The software worked! Their brains barely did.

Nobody outside Facebook knew what was coming, except the few partners that had agreed to develop applications in advance. The company had kept the purpose of f8 secret. The only thing most of Silicon Valley knew was that Facebook would make a big announcement. Facebook had never done anything like this before. Hundreds of journalists crowded the front rows. It seemed every software and Internet company in California, and many from farther away, sent a delegation.

As f8 began, the 750 people throughout the packed room strained to see the diminutive Zuckerberg, wearing his standard T-shirt, fleece jacket, and sandals. He walked out onstage and pronounced, "Together, we're starting a movement!" It was a phrase suggested by hip San Francisco strategy and marketing consulting firm Stone Yamashita.

Zuckerberg's platform demo was by far the best-rehearsed presentation he had ever given. He'd slaved over his wording, but continued modifying his slides until minutes before he was scheduled to appear. He was extremely nervous. Everybody would be watching, even his parents, who were in the audience. But he paid a price for his last-minute modifications. When he got onstage the slides appeared in the wrong order and his speech got out of synch. He paused and looked confused. The event staff and Facebook's executives held their breath. "Well, this worked in my office . . ." he joked. The tension defused. The correct slide came up. He finished smoothly.

The platform wowed the crowd. It took Facebook way past MySpace. No other consumer website had anything like this. Rapturous coverage instantly began sprouting on blogs and journals all over.

A solid ecosystem had already started coming together. More than forty companies demonstrated applications. Mighty Microsoft showed two apps that helped integrate existing Internet software with Facebook. The *Washington Post* (who else?) showed a "political compass" to compare your political views to those of your friends. Sean Parker teamed up with Zuckerberg's old Harvard dormmate Joe

Green to make an application called Causes, to help nonprofits raise money. Another big partner at the platform launch was iLike, which had previously built its own social network to share songs and musical favorites.

Immediately afterward, f8 turned into an eight-hour public hack-athon, where any developer could work alongside Zuckerberg and Facebook's programmers to build software on the fly. (That was another reason it was called f8.) But when the event ended at twelve, the night was not over for the Facebook crew.

They retired again to the W Hotel, where they proceeded to, as they said, "push the platform live," meaning turn it on. Staffers scattered throughout conference rooms to do various necessary tasks, while Moskovitz and Morin sat on a sofa in the lobby working from their laptops via the hotel Wi-Fi. Once the platform was working they crashed, though not before a little partying, of course.

Dave Morin awoke blearily the following morning to find a string of panicked messages on his cell phone. "We have so much traffic we don't know what to do!" said one from an executive at iLike. "Can you help us get more servers?" Apparently just about every application launched the day before was having trouble under the strain of a massive influx of users. Morin headed developer relations, so the companies wanted his help. iLike's executives flew down from Seattle and rented a U-Haul truck, which they drove around Silicon Valley borrowing servers from various tech companies so they could handle the load. By Friday, the day after f8, 40,000 Facebook users had installed the iLike application. Two days later, the figure had soared to 400,000.

Morin got help from the company that ran Facebook's South San Francisco data center. Facebook itself occupied a series of what are called "cages"—fenced-in indoor enclosures full of servers and networking equipment. An adjacent cage was made available to any developer that needed help managing its traffic. Eventually Facebook did a deal with a larger data center operator to open an entire facility for application partners, which would be, in Internet lingo, "peered" with

Facebook's, meaning that in the electronic topography of the Net it was essentially right next door.

The reaction to f8 across the tech industry was close to ecstatic. Facebook's platform launch became—along with the launch of Apple's iPhone a month later—one of the two most-discussed tech events of the year. No longer was it possible to dismiss this upstart as merely a plaything for college kids. The influential blog TechCrunch called the platform "inspired thinking." Prior to f8, Zuckerberg and his crew had hoped that in the subsequent year 5,000 applications might come onto Facebook and half its users would install them. But within six months 250,000 developers were registered, operating 25,000 applications.

Just as Zuckerberg had predicted, Facebook gave applications an unusual ability to acquire new users. This was the vaunted "distribution." The News Feed told users when their friends had installed new applications, so even the most modest app from a single developer with no marketing budget could reach millions of users almost overnight if it did something useful. Though the News Feed still was a selection chosen by algorithm, Facebook tuned the software to make sure that newly installed applications were announced. By six months later, half of Facebook's users had at least one application on their profile.

Just about every software and Internet company was suddenly talking about building an application for Facebook—from industry titans to college kids in their dorm room. Facebook's platform infrastructure made it almost as easy for such lone wolves to create an application as for Microsoft. When it launched the platform, Facebook turned off its own Courses application, which helped college students track one another's class schedules. A New Jersey high school student named Jake Jarvis, seeing opportunity, quickly wrote something similar and six months later sold it for an amount his father says was "sufficient to pay for a year in college."

The platform brought Facebook a gravitas it never before possessed. It caused both technologists and ordinary users to sense that this service was more than they'd reckoned. In Silicon Valley and among techies worldwide, it suddenly became uncool not to have your own Facebook profile.

The platform also changed the experience of being on Facebook. There was a new expansiveness, an air of possibility. If adding the photos application had made Facebook feel like a place where you wanted to spend a lot of your time, turning it into a platform for applications began to make it feel a bit like being on the Web itself. Facebook was becoming its own self-contained universe.

For high school and college students it had long been routine to spend the majority of their online time there. Now people of all sorts and of all ages began to do the same. On the day of f8—May 24, 2007— Facebook had 24 million active users, with 150,000 new ones joining every day. The demographics were already spreading out, with 5 million users between twenty-five and thirty-four, a million between thirty-five and forty-four, and 200,000 over age sixty-five. Within a year Facebook tripled to more than 70 million active users.

In all the complex and frenzied preparations for f8, Zuckerberg and his team had given surprisingly little thought to exactly what kinds of applications were likely to work best on Facebook. As is so often the case at this company, driven by ideals and led by a CEO obsessed with a long-term view, high-mindedness prevailed. The Facebook team assumed that general-purpose applications with wide functional appeal would play a big role in the new ecosystem. When they prepared for f8 by taking proprietary features out of their own photos app, for example, they believed that someone might come along with a better one and successfully compete against them. Their idea was that this should be a forum for the best, most functional, most sophisticated applications. When I was reporting in 2007 prior to f8, Facebook had me speak with one close ally of the company, who told me, "Facebook is creating the opportunity to build a whole generation of Adobes and Electronic Arts and Intuits that live within Facebook." These were the giants of the industry. As usual, the company was aiming high.

Facebook, however, is nothing more than the collective actions of its users. What happens there depends on what Facebook users are interested in, not, in the end, what Mark Zuckerberg thinks they ought to be interested in. With Facebook's platform, he learned that lesson a bit painfully.

A frenzy of new applications quickly emerged on Facebook, but they were hardly high-minded. The ones that took off fastest were mostly silly, but intrinsically social in a way that games on the Web had never been before. One of the first really hot apps was one called Fluff Friends. It didn't do much more than let you electronically "pet" a virtual dog or cat, but when you pet your friend's dog your photo would show up on their profile. It was a new way to send a simple message, which 5 million people did. Another similar app enabled you to give your friends a "vampire bite." Food Fight helped you throw food at your friends, and reached 2 million users in just a few weeks. A silly little app called Graffiti—which let you scribble on friends' pages—became the number-two application. A couple of young guys in San Francisco wrote it in a couple days in their apartment.

These were genuinely social applications—they successfully brought offline behavior into this new online world. It's just that it was the kind of behavior that reflected the penchants of the people who still were the overwhelming majority of Facebook's users—teenagers and college kids.

A few weeks before f8, Morin had coffee with Mark Pincus, the erstwhile founder of Tribe.net, co-owner of the sixdegrees social networking patent, and early investor in Facebook. Pincus told Morin excitedly that he intended to build a poker application for the new platform. "It won't work," Morin asserted dourly. "Games aren't viral." Pincus went ahead and launched Texas HoldEm Poker on Facebook, starting a company called Zynga, which was headed for huge success. Zuckerberg himself was disappointed at the silliness of many of these apps. He wanted his company to help people communicate things that mattered, not make it easier to play around.

Then came the phenomenon called Scrabulous. Two brothers from Kolkata, India, Rajat and Jayant Agarwalla, built a blatant imitation of the classic board game Scrabble for Facebook. You could play multiple games with as many friends as you wanted, taking a turn whenever it was convenient. Scrabulous was a sensation. The magazine *PC World* rated Scrabulous number fifteen on its list of "The 100 Best Products of 2008," just behind Craigslist and ahead of the Nintendo Wii. (Facebook itself was number three.) Some days as many as 342,000 people played.

Scrabulous got even Mark Zuckerberg's attention. He had never succeeded in convincing his grandparents to join Facebook, but now they finally agreed—so they could play Scrabulous with him there. His antipathy toward games on Facebook began to crack. It was apparent to him that you were interacting with people you cared about when you played Scrabulous. And after all, Scrabble was a game of words and the intellect—the kind of game played by people who went to Harvard.

All this excitement did not go over well with the owners of Scrabble, however. Shortly after Scrabulous launched, Hasbro, which owns the rights in the United States and Canada, tried to buy the online game, reportedly for as much as $10 million. The Agarwalla brothers refused to sell, then Hasbro sued them. The game was shut down. Meanwhile, Mattel, which sells Scrabble in the rest of the world, launched its own Facebook version for use outside North America. Eventually Hasbro launched a legal U.S. Scrabble for Facebook, and the Agarwallas reworked their game to resemble Scrabble less and renamed it Lexulous. It remains popular.

Pincus's Texas HoldEm was the next game to take off. Morin was right in a sense—games didn't spread virally as quickly as some kinds of applications. But they engendered extraordinary loyalty—once a user starts playing, they frequently come back. Zynga later added other games, including Farmville and Mafia Wars, both of which now have millions of users. Pincus raised money from venture capitalists and invested aggressively. Zynga is now the largest application company on Facebook, with about 250 employees and over $200 million in annual revenues. And Pincus says Zynga is profitable. Texas HoldEm had 20.3 million active users on Facebook in December 2009, making it by far the most popular poker site anywhere on the Internet. But even more impressive is the game called Farmville, also created by Zynga. On Farmville, a player manages and grows a farm, tending crops and feeding animals, etc. You trade with your neighbors and join in a community of farmers all trying to build the biggest and most productive farm. It has about 80 million total users. Overall Zynga has 241 million total active users for all its games as of February 2010, according to the research firm Inside Network.

Games are now the most successful type of application on Facebook, drawing in phenomenal numbers of players. It makes sense, since gaming is a fundamentally social activity. Facebook enables you to play any game with any of your friends on the service. As of February 2010 there were twelve games on Facebook with more than 20 million players, according to the company. The highly complex World of Warcraft for years dominated online multiplayer gaming, with at most 11.5 million players. But gaming on Facebook is a more casual activity. "We now have tens of millions of people playing games who don't identify themselves as gamers," says Gareth Davis, who oversees the gaming portion of the Facebook platform. "They play games here because they want to have fun with their friends."

Davis is working with every major console game manufacturer to enable classic video games to connect with Facebook and incorporate a social element. "In three years every game will be social," he predicts. "Every single device—whether it's a console or a phone or a TV—will connect with Facebook and be able to incorporate and share your Facebook data." One game by a company called Social Gaming Network enables people to play tennis. For a racket they swing an iPhone that is connected to Facebook. Their adversary on Facebook can be anywhere else in the world, swinging his or her own iPhone.

Games and silly applications continued to surge throughout the Facebook platform's first year, but the company was finding it wasn't simple to manage and police its ecosystem of partners. Since anyone could create an app, the platform attracted quite a few players who were less idealistic than Zuckerberg and more interested in making a quick buck.

A competition ensued among applications for users at all costs. Applications were designed as much to get new users as to be fun or valuable. The key was to figure out how to manipulate Facebook's software so that messages would go into people's News Feed inviting them to download an app. Applications became very clever about generating stories that would flood everyone's home pages. One called Funwall let you create little animations or download video onto your profile. That was fine and good. But it had an insidious interface that used ambiguous language to trick many users into sending invitations to every single friend. Even tech industry sophisticates fell for it.

Facebook kept trying to weed out spammers and encourage more reputable applications. But changes intended to punish malfeasance often impaired legitimate apps. Says Morin: "We had to learn a lot about developer relations and policy setting and things that we just didn't understand. We sort of stumbled our way through becoming good at dealing with developers."

The company implemented a variety of new rules to try to police applications and make them behave. It urged users to complain about spam. It changed the software to reduce the number of application stories flowing into a user's News Feed. And it hired an industry veteran to head up the platform. Ben Ling, a slender and flamboyant Chinese-American, had been running the payment system called Google Checkout. He was the highest-level employee Facebook had ever lured away from Google. Executives called him a "rock star."

By the summer of 2008 the problems had gotten completely out of hand. Facebook's platform was like the Wild West. So at a second f8 that July, Facebook announced a variety of refinements and rule changes, including a rating system. Now Facebook could weed out apps by "verifying" the good ones. Facebook wanted to encourage the apps that were the most fun or useful. Despite all the fluff, a fair number of substantial and useful applications did get traction. A popular one called Visual Bookshelf let you list books you've read, rate them, and write short reviews.

But Zuckerberg's favorite Facebook application was the Parker-and-Green–created Causes. It was driven by high motives—to help nonprofits raise money. Facebook users who make a donation create a story in their friends' News Feeds. Ideally that inspires friends to make their own donations. Explains Joe Green: "Social recognition matters in charity, too. People who make big donations like to get their names on hospital buildings. So we at Causes allow you to show what you care about on your Facebook profile." He says it's like wearing a yellow rubber Livestrong bracelet. Users responded strongly. Causes has remained among the largest applications on Facebook.

Now the platform ecosystem has become substantial. There are more than 500,000 applications operating on Facebook, created by over 1 million registered developers from 180 countries. More than 250 of these applications have at least one million active users every month. Investors have high hopes for this new type of software company. The top five Facebook applications companies alone—Zynga, Playfish, Rock You!, CrowdStar, and Causes—have raised approximately $359 million in investment capital between them. That includes a giant $180 million infusion into Zynga in late 2009 by private investors led by the Russian firm Digital Sky Technologies. Justin Smith, who runs Inside Facebook, which is devoted to the Facebook developer community, estimates that there are about fifty venture-funded software companies with substantial revenues whose primary business is building and operating applications on Facebook. Zynga is the largest. About 200 smaller companies, comprised of two to four developers each, have annual revenues of several hundred thousand dollars. At least another 300 solo operators have written a Facebook application that earns enough to support them.

Facebook application companies are doing so well that their estimated aggregate revenue in 2009 was roughly the same amount as Facebook's itself—slightly over $500 million. These applications generate revenue in several ways. Selling advertising generates $200 million for applications companies. Apps often host ads promoting other Facebook apps, and get paid about fifty cents on average each time a user clicks through and installs another app.

Transactions inside applications create even more revenue. Justin Smith of Inside Facebook estimates there were $300 million in such transactions in 2009. Much of this is spent to buy an upgrade to a more advanced level of a game, or to buy virtual goods, like a fancier shoe to kick your friend in KickMania. Playfish's Pet Society game, where users set up houses to display their pets, releases new virtual items every Monday. On Valentine's Day in 2009 the company sold five million images of roses that players could give to their friends. Each one cost about two dollars. In Zynga's Texas HoldEm, players who want more chips than they're allocated each day pay real money to get them, even though there is no way to remove winnings from Facebook. Numerous Facebook games have revenue exceeding $3 million per month.

Savvy marketers have also realized that Facebook applications are a free way to get in front of consumers. That's why the Washington Post Company did its Political Compass. When Bob Dylan released a new album in 2008, his record label created an application that used old footage of him as a young man holding a series of signs. Facebook users could put their own message on the signs and then host the film on their profile.

In turning its network into a platform for whatever any outside developer wants to build, Facebook has created many new capabilities but also a new set of risks. For all their usefulness and entertainment value, applications on Facebook are often cavalier about how they treat user data. Frequently when users install an app they give it essentially blanket permission to extract data from their profile. But once that data is in the hands of the developer the user loses all control of what happens to it. Facebook has just begun to take steps to deal with this problem. The limits for what is and isn't acceptable remain unclear, and predatory applications continue to arise that take unnecessary liberties, often in order to make personal data available to outside marketers who pay for access to it. It's another piece in the complicated puzzle of Facebook privacy. Says Marc Rotenberg of the Electronic Privacy Information Center: "Facebook and its business partners learn lots about us, but we know very little about them or about what information of ours is collected and how it's used."

As more and more software companies embrace the platform and as Facebook's dominance of social network computing spreads around the globe, the company's platform strategy is rapidly evolving. Its long-term plan is that fewer and fewer applications will operate inside Facebook's own walls. Now a service called Facebook Connect enables any website to tap into users' information and network of friends, and send reports of user activity back into News Feeds on Facebook. The company increasingly is encouraging partners to tap into Facebook that way. So far more than 80,000 websites already do, including about half of the largest ones worldwide. Zuckerberg's long-desired platform strategy has been paying off.

12

$15 Billion   "A trusted referral is the holy grail of advertising."

Opening Facebook up to everybody had been a huge success. By the fall of 2007 more than half the site's users were outside the United States. The explosive international growth was a powerful sign of Facebook's growing universal appeal, since the company had done nothing to make it easy for non-Americans to join. All the text remained completely in English, for one thing.

But the growth also presented a serious business problem. Facebook had to start figuring out how to make money—providing service to people all over the world was expensive. All the ads were aimed solely at Americans, which meant that Facebook was not generating any appreciable revenue from more than half its users. The advertising deal Facebook had signed with Microsoft a year earlier only applied inside the United States. If Facebook was going to take advantage of its newfound global presence, it needed a partner to help it sell display advertising internationally. Microsoft had made it plain it would love to be that partner, extending its U.S. deal to a global one.

Zuckerberg had always been blasé about advertising . But Facebook now had 50 million active users and the platform had transformed it into an industry darling. The company had to find a way to pay for it all. Hundreds of thousands of new users were joining every week. And Facebook continued to build its infrastructure based on the assumption it would be much, much larger in the future. That meant spending millions of dollars for new servers. If he had to have ads, Zuckerberg hoped to develop a new kind that would work uniquely well on Facebook, ads that wouldn't interfere with a user's experience. The last thing Zuckerberg wanted was for it to feel like watching network television, where the show is routinely interrupted by irrelevant and inane advertising.

The U.S. ad deal gave Microsoft the exclusive right to sell banner advertising on Facebook. That had to change. Relying primarily on Microsoft for the lion's share of revenue was precarious. Facebook needed its own self-managed streams of revenue.

Separately, Zuckerberg and Facebook's board decided it was time to raise more money. Peter Thiel in particular wanted to do it that fall. Thiel has a refined nose for the twists and turns of the financial markets. Stocks were at levels not seen since the dot-com bubble and investors were feeling buoyant. Facebook's reputation and growth had been transformed by f8 and the platform launch, and it was time to take advantage of enthusiasm among investors. But Thiel also knew that if they went out asking investors for money, someone might try to buy the entire company. To Thiel and Jim Breyer that was an appealing idea, but it horrified Zuckerberg (whose two unfilled seats on the board of directors remained in his control as a form of horror insurance).

The CEO asked Van Natta and his newly hired chief financial officer, Gideon Yu (who had been CFO at YouTube), to see what kind of interest they could drum up for a small stake in the company. Yu now says he thought Facebook might be able to get investors to buy stock at a valuation of about $4 billion. That would have been a huge leap. A little more than a year earlier, its third round of financing (called Series C) had raised $27.5 million, valuing Facebook at $525 million.

But this company, as usual, didn't conform to the usual expectations. Several venture capital and private equity firms were willing to buy a chunk of Facebook at a $10 billion valuation. That was eye-opening to Yu. He'd clearly been thinking too small. But Zuckerberg wasn't satisfied. He thought the company was worth $20 billion, says another confidant. He and Van Natta decided to try for $15 billion. Sure enough, they found interest from several parties, but not enthusiasm. Nobody would invest at this level without doing some serious negotiating on the terms. "We found where the market was," says Yu. "We were going to be able to get a deal done at $15 billion."

This was right about the same time that talks with Microsoft about an international ad deal began in earnest. The software giant also wanted to hold on to its U.S. deal with Facebook, but Microsoft executives felt they couldn't let it stay the same. It needed to renegotiate the deal as much as Facebook did. Microsoft was losing about $3 million a month on the U.S. ads. It was putting the lion's share of its banner ads on Facebook pages that displayed photos, but people just didn't much notice ads in that environment. The price Microsoft thus could charge advertisers was low. Yet it had agreed to pay Facebook a guaranteed minimum amount—around 30 cents for every thousand page views, regardless of what it was getting from advertisers.

Everything Microsoft was doing in online advertising was fundamentally a response to Google's growing power. Google search ads were garnering over half of all online advertising dollars, even as the increasingly profitable search giant was starting to toy with other sorts of software that directly competed with Microsoft's core PC products. To fight back and defend its turf, Microsoft was now going head to head with Google across the board in online advertising. As part of this effort, it was investing billions of dollars to improve its own online search software. Separately, it had in May made its biggest purchase ever—paying $6 billion for aQuantive, which distributed advertising across the Internet. Now that it owned this distribution engine, it badly needed additional inventory to sell through it.

Microsoft CEO Steve Ballmer was fed up with losing deals to Google. He had recently lost both of the industry's two biggest partnership opportunities after coming exquisitely close to agreement. Each time, Google swooped in at the last minute and stole the deal away. Ballmer had flown to New York in December 2005 to negotiate a major ad partnership with Time Warner's AOL. He left town thinking he had a deal. Google unleashed its ad team headed by Tim Armstrong and came in with a better offer in days and sealed a contract with a $1 billion investment in AOL that valued it at $20 billion. Then in August 2006, Microsoft had a deal with News Corp.'s MySpace and was poised to guarantee $1.15 billion, according to one of the deal negotiators. Google pounced at the final hour and won with a three-year guarantee to News Corp. totaling about $900 million. News Corp. apparently wanted the status of partnering with Google so much it was willing to give up revenue to do it. Microsoft had been further galled when Google swept up Internet banner advertising network DoubleClick for $3.1 billion in early 2007. This time, Ballmer was resolved it would not happen again.

Van Natta's forte is deal-making. He coolly played Microsoft off against its archrival. He knew that uttering the word "Google" was like a magic spell to tame Microsoft's normally rapacious negotiating instincts. And in fact when Google heard that Facebook was looking for a partner for its international ads, it began aggressively pursuing a deal itself.

On October 10, 2007, Google hosted its signature annual event for its best advertising clients—called Google Zeitgeist. Not only did the biggest marketers and ad agencies come to its campus for the two-day conference, but Google's board of directors converged too for one of their quarterly meetings. It was a good time to make deals. Tim Armstrong, Google's ad chief, had talked to Van Natta and knew that Microsoft was well along in talks to win Facebook's international ad contract. But it just so happened that Mark Zuckerberg was one of the signature speakers at Zeitgeist. Armstrong talked to Google's board members and got an official go-ahead to use this opportunity to begin serious negotiations with Facebook to try to take the deal away from Microsoft. The board even approved talks about buying Facebook, if it made sense.

Google didn't make any secret of its interest in an ad deal with Facebook. At a press conference during Zeitgeist, Google CEO Eric Schmidt called social networking "a very real phenomenon." He added, "People don't appreciate how many page views on the Internet are in social networks." It was an early expression of what would become a long-standing concern—Google cannot search content that is behind proprietary walls on the Internet. A close relationship with Facebook might achieve even more than acquiring a bunch of new ad space. It could help Google stay dominant as the Internet evolved.

That evening everybody was bused to a nearby park where Google had erected a gigantic white tent. After a lengthy cocktail hour, all 250 or so Zeitgeist attendees sat down to a lavish, almost Bacchanalian feast. The first course was served on thick plates made of ice. Google was at the height of its powers—money was flowing in like manna. Here the company was thanking the people who were spending those billions on advertising while simultaneously proclaiming itself to be rich rich rich. At the center table, immersed in intense conversation, were Google co-founder Larry Page, Armstrong, deal expert Megan Smith, as well as Zuckerberg, Van Natta, and Facebook corporate development boss Dan Rose. The Facebook team said they were far along in negotiations with Microsoft. Armstrong impressed upon the Facebook executives that Google was serious about wanting to do the deal instead.

After the lavish dinner ended at around 10 P.M., Facebook's trio and the Google executives retired to the company's nearby headquarters building for some serious negotiating. They worked late into the night until they had the rough outlines of a deal. Google would take over both the U.S. and international ad deals. It also agreed to consider making a small investment in Facebook at the $15 billion level. For Google it made sense to buy the equity as a sweetener because Facebook would have to go through quite a bit of trouble to dislodge Microsoft. If Facebook pushed Microsoft out of the still-in-effect U.S. deal it was likely to result in legal unpleasantness. But Google had gone further. Executives told Zuckerberg that they were willing to consider buying Facebook outright, though at a price considerably less than $15 billion. This time, however, Zuckerberg was firm. Facebook was not for sale.

Even on the ad agreement, many on the Google side detected a lack of commitment on Zuckerberg's part. They noticed he kept pushing for very specific concessions on things like the size and shape of display ads—the kind of thing usually left to underlings to iron out. It seemed to them he might be seeking specific promises from Google in order to strong-arm Microsoft to concede the same points. For all the talk, the Google team knew that Microsoft's prior relationship with Facebook gave it a big advantage. The chances of pulling Facebook away remained small.

Microsoft had been carefully cultivating Zuckerberg. CEO Steve Ballmer had flown to Palo Alto to visit his young counterpart twice. Ray Ozzie, Microsoft's Chief Software Architect, had also repeatedly visited Palo Alto. As Zuckerberg is wont to do, he took them on long walks. He told Ballmer that Facebook was raising money at a $15 billion valuation.

But Ballmer had come with something very specific in mind. "Why don't we just buy you for $15 billion?" he replied, according to a very knowledgeable source. Zuckerberg, as usual, was unimpressed, even by this fabulously extravagant offer. It was so high that, had it been accepted, Microsoft's shareholders might have raised serious obstacles to its completion.

"I don't want to sell the company unless I can keep control," said Zuckerberg, as he always did in such situations. He knew that keeping control once Facebook sold would be almost impossible, so for him this was effectively a way to end the conversation.

Ballmer took this reply as a sort of challenge. He was emphatic that Microsoft wanted to buy Facebook. He went back to Microsoft's Redmond headquarters and concocted a complicated plan intended to begin a process of acquisition without compromising Zuckerberg's ability to call the shots. According to people close to the situation, Ballmer proposed that Microsoft acquire a minority stake in Facebook at a $15 billion valuation. Then, in a provision loosely modeled on a deal arrived at almost two decades earlier between giant Swiss pharmaceutical firm Hoffman-LaRoche and Silicon Valley biotech star Genentech, Microsoft would have the option, every six months, to buy another 5 percent of Facebook. A complete takeover of the company would take 5 to 7 years, depending on how much of the company Microsoft bought at the outset. The price Microsoft was obligated to pay would rise steadily over time, making Facebook's ultimate price considerably higher than $15 billion. But from Ballmer's point of view, the deal addressed Zuckerberg's key concern—it allowed him to retain control, at least for another few years.

Ballmer flew down to San Francisco again and brought along Kevin Johnson, who oversaw all Microsoft's ad-related business. Van Natta suggested that they all meet at his Palo Alto home in order to avoid attracting attention. Ballmer didn't make that any easier. He arrived in a big black Cadillac Escalade with a contingent of security personnel wearing earpieces and microphones. As the security men cased the yard, Van Natta delivered bad news to Ballmer and Johnson, while Zuckerberg sat quietly and Microsoft software chief Ray Ozzie listened in via speakerphone.

Facebook wanted to alter the U.S. deal, Van Natta declared. In fact, it was going to start selling its own ads soon, whether Microsoft liked it or not. Ballmer was nonplussed. If Microsoft wanted the international deal, Van Natta continued, it had to agree to concessions on the U.S. one. Facebook needed to try out some new ad formats of its own. If Microsoft wouldn't agree, well, Google was waiting in the wings.

The encounter of Van Natta and Ballmer must have been a sight to see. Van Natta may be fearless, a belligerent and uncompromising negotiator, but Ballmer is big, loud, and consummately forceful himself. You don't screw around with him. Not to mention he's CEO of what is in financial terms still the most powerful technology company in the world. Van Natta must have had an exquisite sense of just how far to push, because Ballmer didn't lose his cool. He reiterated that Microsoft had no interest in reopening the U.S. ad deal. What he was really interested in, he said, was buying Facebook.

Zuckerberg was cautious. The young CEO had learned his lesson the year before with Yahoo—once you open the door to a possible sale it's hard to close it. Zuckerberg made it clear he was inclined not to sell, but suggested that Microsoft would have to agree to a raft of conditions, including even more autonomy for Facebook than the tiered proposal anticipated, with Zuckerberg remaining at the helm indefinitely. Says someone from Microsoft who heard about what happened, referring to Zuckerberg: "It wasn't 'If you pay $X billion we'll do it.' The guy's not a seller. His expectations were too high." Despite all the work Ballmer had put into his acquisition proposal, Zuckerberg wouldn't bite.

While Microsoft was almost desperately seeking the international deal, Facebook took advantage of the software giant's pliability to resolve another dispute, a problem with Hotmail, Microsoft's free Internet email service. The biggest tool for Facebook's growth was the contact importer it had launched at the time of open registration. New users entered their email username and password, and Facebook helped them send anyone on their email list an invitation to join. Hotmail was the largest source by far of such user referrals. But it interpreted many email invitations coming in from Facebook as spam—unwanted commercial messages. On some days Hotmail blocked the use of the contact importer altogether. Facebook's user growth then dropped as much as 70 percent, says Moskovitz. So in the midst of the ad talks, uber-negotiator Van Natta, Moskovitz, and D'Angelo trooped up to Microsoft headquarters in Redmond, Washington, to iron out the conflict. "This was absolutely not something we could walk away from," says Moskovitz. After a day or so of talks, Van Natta got Microsoft to stop interfering with the imports even though Facebook conceded almost nothing in return.

In a classic brinksmanship move, Microsoft's executives told Van Natta they wouldn't budge on letting Facebook sell some of the U.S. ads. Van Natta refused to release Microsoft from its minimum payments for ads next to photos. Microsoft ad chief Johnson replied that keeping the U.S. inventory was critically important to him. In order to keep it, he said, he was willing to lose the international deal. "Fine," said Van Natta. "We're going with Google."

Johnson returned to Microsoft's headquarters in Redmond, Washington. But Hank Vigil, the company's top dealmaker, stayed in Palo Alto to continue talks with Van Natta. He made a breakthrough. Van Natta had told the Microsoft team that his job depended on a successful outcome of the talks. Now he waffled. Vigil proposed that Microsoft would let Facebook use 15 percent of the U.S. ad inventory if it made certain concessions. On Saturday morning Vigil set up a conference call with Kevin Johnson and Van Natta to expand on the offer. Johnson said he'd agree to release the inventory if Microsoft got the international deal and if Facebook eliminated the photos ad minimums and agreed to use Microsoft's search engine inside Facebook. Johnson instructed Vigil to take his team to Facebook's offices on Monday morning and not leave until he had a deal.

By 11 A.M. Monday, all the players were ensconced in a conference room on the second floor of Facebook's office on University Avenue in Palo Alto. Sitting at a big glass table, with sunshine streaming in through the walls of glass, the two companies' teams conducted the top-secret negotiation that would transform Facebook's reputation. In another Facebook building, a smaller group from Google met for part of this time, discussing its own possible deal.

For the next twelve hours the Microsoft and Facebook teams went back and forth on issues large and small. Microsoft got an agreement to move toward providing search technology inside Facebook, yet another blow in its near-feudal joust with Google. Facebook demanded Microsoft not display ad banners either at the top of the screen or on the lower left, just on the lower right side. (Google's acquiescence on this point gave Facebook ammunition.) Unlike the year-earlier U.S. deal, there would no longer be any up-front guarantee of how many ads Microsoft would display nor how much it had to pay Facebook. The two companies would instead share the revenues from any ads sold. Facebook successfully pushed for a higher percentage than is usual in such deals. And the social network got its all-important flexibility to experiment and innovate on new ad formats on 15 percent of the U.S. display ads.

The Facebook team—Van Natta, Rose, Yu, and General Counsel Rudy Gadre—occasionally ducked around the corner to huddle with Zuckerberg, whose desk was only steps away. The CEO was much more involved than in past such negotiations. Whenever things bogged down, Van Natta loosened Microsoft up again with a vague allusion to Google. He implied, but didn't exactly say, that Google was ready to do all the things Microsoft wouldn't. It was close to true, anyway.

At about 11 P.M. it was apparent a deal was in sight, though many details remained to be resolved. Everyone's energy was flagging. Just then, a blast of thumping house music invaded the quiet conference room. Several negotiators stepped out into the loftlike office to see what was going on. They found a Facebook programmer at a DJ stand, with the music turned all the way up. This was the signal to Facebook's engineers that a hackathon was about to begin. These were the all-night sessions, legendary in Facebook's engineering culture, when many of the site's most interesting innovations emerged. Unlike a typical hackathon, though, this so-called "convertathon" had a specific agenda—to change Facebook's underlying software code to make it easier to translate into languages other than English. Translation of the site was set to begin in a few months in order to further bolster already torrid international growth.

Back in the conference room, heads began to bob and feet to tap. It was ludicrous, but energizing. A Microsoft negotiator got up and stood cordially in line with Facebook programmers, waiting to dig into the bins of take-out Chinese food. Negotiations picked back up once the music was turned down. By 3 A.M., they had a deal. Essentially Facebook got everything it wanted. The negotiators left the engineers to their labors and went to bed.

The issue of an accompanying investment had not been a topic during the glass-room negotiations, but the next morning, Van Natta raised it bluntly. Says his then-deputy Rose: "We said to them, 'Look, if you want to use the investment opportunity to cement the relationship, we want you to lead the round.' We said, 'We might be talking to your competitors.'" Microsoft had continued to make clear that if Facebook was willing to sell, it was interested in buying. But Zuckerberg was not about to sell. Van Natta was prodding Microsoft to buy a small chunk. Ballmer had in effect already agreed that Facebook was worth $15 billion, so the valuation wasn't much in dispute. At that astronomical level even a small percentage of the company would net Facebook many millions that it could use to underwrite its money-losing operations. The Hong Kong billionaire Li Ka-shing, often called "Asia's Warren Buffett," had earlier approached Facebook to invest and had been negotiating hard during these very same days and had agreed to invest at that valuation. "It was a frenzied period," says Yu. Everybody was acting as if Facebook could become a financial colossus, even though at the moment the only thing huge about it was its membership growth rate.

After some frenzied back-and-forth occurring mostly over the course of a single day, Microsoft agreed to invest $240 million at a $15 billion valuation for 1.6 percent of Facebook, alongside Li Ka-shing, who would put in $60 million for 0.4 percent. Microsoft's executives were happy. "It was all about the search war with Google," says one. "A $240 million investment that helped us fight them was definitely worth it." The pressure was great to conclude the deal, so Microsoft had little time for any detailed financial due diligence. But it was critical to Microsoft that another investor participate alongside them in the round. Microsoft had to be able to demonstrate it wasn't paying an inflated price in order to achieve an ad deal. Otherwise, were Facebook later to be determined to be worth less than $15 billion, accounting rules would require Microsoft to write down as a loss on its books the proportional difference between $15 billion and the actual valuation. So Li's stake, while small, was crucial.

Microsoft did not get a particularly attractive deal in legal terms, either. In order to move quickly for this so-called Series D round, it agreed to abide by the same documents that had applied to investors in the Series C when several VCs had invested in mid-2006. The convertible preferred shares it bought had what is called a "1X nonparticipation liquidation preference," which means if Facebook were ever sold outright, Microsoft would get back either its actual cash outlay of $240 million or 1.6 percent of the purchase price, whichever was larger. But it could do nothing to stop a subsequent investment round at a lower valuation. If there is eventually a public offering of Facebook's stock, Microsoft will be forced to convert its preferred shares into common stock proportionate to its ownership, no matter how much the company is then worth, whether more or less than $15 billion. Microsoft was willing to accept all these conditions because its primary goal was the completion of the advertising deal. But at the last minute it demanded one important condition: Facebook could not take any investment money from Google. And if it ever contemplated an outright sale to the search nemesis, Microsoft had to get advance notice.

The deal was announced on Wednesday, October 24, and prompted an outcry of amazement. The *Wall Street Journal* called Facebook "the newest Internet darling" and said the deal was "reminiscent of the Internet bubble that ended in 2000." The *Los Angeles Times* called the $15 billion figure "staggering." "It tips the scales in terms of totally ridiculous valuations," wrote the influential TechDirt blog. This was by far the highest valuation ever given to a private technology company, and one with no profits to boot! Either Microsoft's Steve Ballmer was insane, or Facebook mattered more than anyone had realized. But if the f8 platform event five months earlier had firmly put Facebook once and for all onto the technology industry map, this investment did the same thing for Facebook on Wall Street. Microsoft's stock jumped markedly.

The ad deal that precipitated the investment was barely noticed in the hubbub over the valuation.

Facebook's timing on the deal could not have been better. Only two weeks earlier, the stock market peaked at a level it has not approached since. In 2008 the world fell into the worst recession of the postwar period. But Zuckerberg had a lot of crucial cash in hand to help him through the down times. In addition to the initial $300 million it raised in Series D, Li Ka-shing invested an additional $60 million several months later, and three Munich-based venture capitalists, the Samwer brothers, invested $15 million around the same time, bringing the total raised in the Series D round to $375 million. Zuckerberg has a simple explanation for how Facebook achieved such an amazing financial result. "Peter [Thiel] helped us time it," he says simply. "He was like 'Now would be a good time to raise money.'"

Now that Microsoft was no longer an obstacle to Facebook selling ads on its own site, Zuckerberg and company lost no time launching a new sort of ad on the service. Only two weeks after closing the Microsoft deal, Facebook hosted its first-ever big event for the advertising community in New York on November 6. The announcement had several parts. Any commercial entity could now create a "page" on Facebook for free, which would have many of the characteristics of an individual's profile, including the ability to host applications. The "sponsored page" model had outlived its usefulness. The company's strategy was to get as many companies into its system as possible, on the presumption that once they were operating there they would find cause to advertise or otherwise spend money, even if their page itself were free.

A user could become a "fan" of one of these pages rather than a "friend," as they did with an individual. Activities of users on these new commercial pages would be broadcast to their Facebook friends' News Feeds. (I soon became a fan of the *New York Times* page, for example, and my friends saw an announcement of that in their News Feeds.) Barely mentioned was the fact that a service called Beacon would also enable forty-four companies, and more later, to extend a similar alert system onto their external websites. Activity on these external websites could go into friends' Facebook News Feeds as well.

The meat of the Facebook Ads announcement, at least in the minds of those who planned it, was that Facebook would launch a new kind of self-service advertising that enabled any company, even a tiny one or an individual, to go online and design and purchase an ad on Facebook that they could target very exactly to their intended audience. In effect, the kind of custom targeting that Moskovitz had pioneered three years earlier—for example, when Interscope Records targeted ads for Gwen Stefani's "Hollaback Girl" to cheerleaders—was now coming to the mass market. One assumption was that the owners of the new pages would be heavy users of these ads as a way to promote their presence on Facebook. Another component of the new self-service ads was what the company called "Social Ads," which would pair a paid commercial message with a Facebook user's endorsement.

At the announcement Zuckerberg attracted considerable attention—and derision—with his grandiose introduction. It was the first time he had ever given a big promotional talk outside the confines of Silicon Valley. Hearing him speak, you might think he had gone almost overnight from despising ads to wanting to own the worldwide ad industry. "Once every hundred years," he began, "media changes. The last hundred years have been defined by the mass media. In the next hundred years, information won't be just pushed out to people. It will be shared among the millions of connections people have. . . . Nothing influences people more than a recommendation from a trusted friend. . . . A trusted referral is the Holy Grail of advertising." Unfortunately, since Facebook's original intention was to promote the new self-service ads, which were more about making targeting available to the masses than about trusted recommendations, Zuckerberg's intro was misleading to begin with.

The feature that would come to define Facebook Ads, and to turn November 6 into a day of infamy, was Beacon and the way it worked outside Facebook's walls. Beacon was a poorly designed alert service. It wasn't even an advertising product, since it generated no revenue. It was built by Facebook's platform team, not the ad group. But while it was intended for activities like playing a game or adding a recipe to an online recipe box, it also could be used to announce purchases you made on partner sites. And Facebook had lined up a bunch of commercial partners for it. If you, say, rented a movie at Netflix, bought a pair of shoes at Zappos.com or a movie ticket on Fandango, you could give the website permission to broadcast that fact to your friends back in Facebook via an item into your News Feed. But Beacon was a last-minute add-on to the Ads launch and had barely been tested with users. Its implications were overlooked by Zuckerberg and his executives in days leading up to the launch.

And it had a major design flaw. When you, say, bought your shoes at Zappos, you weren't asked to explicitly approve sending that fact to your friends inside Facebook. Instead, you were shown a little drop-down menu that asked if you wanted to not send the information. If you didn't proactively stop the alert, it would proceed. In Web lingo, that's called "opt-out" rather than "opt-in." And the opt-out menu only displayed for a few seconds before disappearing. Many users seemed to miss it altogether.

After the launch, stories began emerging in the press of users who had unintentionally launched word of their commercial actions back into Facebook with unfortunate consequences. One Massachusetts man bought a ring, whereupon an item appeared in his wife's News Feed: "Sean Lane bought 14k White Gold 1/5 ct Diamond Eternity Flower Ring from overstock.com." Within two hours Lane's surprised wife, Shannon, sent him an instant message: "Who is this ring for?" In fact it was to be her surprise Christmas present, according to a story in the *Washington Post.* Lane told the *Post* that he was "crestfallen" that his surprise was spoiled (and also possibly that Shannon's News Feed item linked to an Overstock Web page showing he got 51 percent off). Another relationship was disrupted when a New York man's girlfriend saw he had purchased a ticket on Fandango to a movie he was scheduled to see with her the following week. A number of users who did lots of shopping at Beacon-affiliated sites found that the entire contents of their Christmas gift list had been published to friends in Facebook.

Beacon felt invasive, and misused personal information. It seemed to many that Facebook wanted to hijack data about its users in order to make a buck. After things started going wrong, many in the press looked back to Zuckerberg's hubris at the introduction as a sort of explanation—Facebook was all about power, and Zuckerberg didn't care what happened to his users. This was a fundamental misreading of the young CEO, but Facebook had become so large so fast that journalists were only beginning to understand it.

The backlash built quickly. As with any Facebook controversy, the viral distribution tools of Facebook itself were well used against it. The liberal political group MoveOn.org stepped in to lead the Beacon protest. It took out ads on Facebook (using the new self-service tool) that asked, "Is Facebook Invading Your Privacy?" It invited users to join a protest group, and 68,000 did. In reality the percentage of users protesting was relatively tiny—0.1 percent versus over 10 percent at the height of the News Feed fracas. But MoveOn got a lot of attention. It and other activist groups were also filing formal complaints with the Federal Trade Commission. Some were preparing lawsuits.

And now whatever happened at Facebook was big news. It had 57 million users and Microsoft's money behind it. The press wanted Zuckerberg to apologize and turn off Beacon. Many writers argued that Facebook's stunning new valuation had made it suddenly, desperately eager to prove it could be profitable. One story that indicates how far Facebook's image had fallen was written by *Fortune*'s Josh Quittner. Titled "RIP Facebook?" it argued the company was "coming undone." Quittner compared twenty-three-year-old Zuckerberg's rash decision making in the Beacon episode to "watching an unattended child play with a pack of matches in a wooden house."

Beacon was the worst and most damaging controversy Facebook has ever faced, for several reasons. First, unlike with the News Feed, the company made a serious product design mistake. Beacon really did result in data being misused. It thus violated Zuckerberg's principles about the importance of privacy and user control of information. But the damage was compounded because for more than three weeks Zuckerberg did nothing to respond to the complaints. As his silence continued, the controversy grew angrier. He was watching user statistics as he always did and saw that Beacon was not affecting behavior inside the service. But that fact belied the genuinely painful experiences of a small number of users as well as the legitimate outrage of the press.

There is an edge of sanctimony to Zuckerberg that at times like this can serve him poorly. But the irony was that he had endlessly resisted, up until now, anything that resembled an intrusive ad or message on Facebook. Here was someone who had said for years that he wanted to do what best served his users, now suddenly acting as if he knew better than they did.

Now, in retrospect, Zuckerberg acknowledges he'd become cocky. "We didn't react quickly enough," he says ruefully, "because we were just so used to people complaining about things and then us eventually being right. We were like 'Hey, whatever, they'll eventually get over it.' Then it was like 'Hey, no, we actually messed this one up.'"

Finally, on November 29, more than three weeks after Beacon debuted, Facebook redesigned it to be a fully opt-in system. No message about you would now be sent without your explicit permission. MoveOn issued a cautious victory statement. Then a week later, Zuckerberg made his first public statement on the mess, with a deeply contrite blog post on Facebook's site titled "Thoughts on Beacon."

"We've made a lot of mistakes building this feature, but we've made even more with how we've handled them," he began. "We simply did a bad job with this release, and I apologize for it. . . . We took too long to decide on the right solution. . . . Facebook has succeeded so far in part because it gives people control over what and how they share information. . . . In order to be a good feature, Beacon also needs to do the same." He also announced that Facebook was now making it possible to turn off Beacon completely, something MoveOn had asked for.

Beacon put a black eye on Facebook that still hasn't fully disappeared. Internally, the Facebook Ads launch was dubbed beforehand with the project name "Pandemic." It really did turn into a disease that was hard to wipe out—a disease of negative perception that lingered long after the product was modified. Membership growth slowed discernibly in the aftermath of all the negative press coverage, though it picked up again by early 2008. Dan Rose, the former Amazon executive who heads marketing for Facebook's advertising efforts and was deeply involved in the Facebook Ads launch, says the controversy was "devastating" for the company. "By the time we fixed Beacon, the meme was already out there that people didn't control the way their information flowed," he says. "We just really screwed it up. It took a long time for our brand to grow past that."

But Beacon did illuminate where Facebook hoped to go in the future—to become a social hub where information about your behavior across the Web was aggregated for friends to see. If you buy something or make a comment on a blog or indicate you like something, Facebook's goal is that eventually it should be possible to let your friends in Facebook know that. In fact, the Beacon program didn't end until late 2009, along with the settlement of a lawsuit about it. Meanwhile, objections to it had died away. Zuckerberg now says he may have simply launched Beacon too soon: "One of the things that was bad about Beacon was that people just weren't ready yet to share their information off of Facebook." The company in 2008 launched a much more widely deployed technology called Facebook Connect for people to share what they do at partner websites. There has been virtually no protest, largely because Connect gives users sufficient control over the information about them that it sends to their friends.

Shortly after the Beacon frenzy died down, board member Jim Breyer had a stern conversation with Zuckerberg. "We blew it," he said. "We should have apologized right away. This, to me, Mark, is an example of why it's so important for us to get a new chief operating officer into the company." Owen Van Natta was a great deal guy, unsurpassed at business development, but not the firm, steady, second hand on the tiller that Breyer thought a company with a still-learning twenty-three-year-old CEO needed. And the company also needed someone well-versed in the complexities of the online advertising business. Zuckerberg took a couple of weeks to think about it, then told Breyer he agreed. He would tell Van Natta of their decision in early January and begin a search.

At a Christmas party in mid-December, Zuckerberg got into a conversation with Sheryl Sandberg. She was a senior executive at Google who had built the search company's self-service ad business into one of the economic powerhouses of the Web. The two of them ended up standing in a corner for over an hour as Zuckerberg queried her about how to manage a growing tech organization. They agreed to get together sometime for dinner.

In the meantime, after a difficult conversation with Van Natta, Zuckerberg began meeting potential candidates for the COO job. One of them was Dan Rosensweig, the former chief operating officer at Yahoo who had only a little more than a year earlier avidly pursued the purchase of Facebook with Yahoo CEO Terry Semel, and who with his wife had hosted the party where Zuckerberg met Sandberg. Another was Jeff Weiner, another top Yahoo executive widely known for his judgment and managerial smarts.

Sandberg, who had been at Google since 2001 and made many millions from her stock options there, had decided she was ready to leave. She had already been offered a great job at a big East Coast media company, and was mulling it seriously. She spent an afternoon talking to Roger McNamee, an industry sage and one of Silicon Valley's best-known investors. She wanted his advice about the media job. "It's a really good idea. You should do this," McNamee told her, then hesitated. "But what you really should do is go work with Mark Zuckerberg at Facebook." McNamee had been informally advising Zuckerberg and knew he was looking for a new COO. Sandberg and Zuckerberg had coincidentally been emailing about a dinner the following week. Sandberg hadn't thought of it as a recruiting dinner, though Zuckerberg's aide-de-camp and in-house recruiter Cohler had talked to her repeatedly for over a year about Facebook. "When are you coming to work with us?" he asked every time he saw her.

By the time Sandberg arrived at the quiet little Silicon Valley restaurant, McNamee had spoken to Zuckerberg and made a case for Sandberg. At dinner the two talked and talked. The restaurant closed at 10 P.M., then Zuckerberg went back to Sandberg's home to keep talking. She's the mother of two small children and she usually goes to bed by 9:30 P.M. By midnight she had to kick him out so she could sleep.

Dinners like that continued. Zuckerberg was in no hurry. He wanted to get to know this person whom he might be working with for the next ten to twenty years. This time he wanted to hire somebody for the long haul. Sandberg says the meetings with Zuckerberg, which he estimates took a total of fifty hours, were "endless." "He never left!" she says in an interview. "Put that in your book. He just would not leave my house."

Sandberg is an elegant, slightly hyper, light-spirited forty-year-old with a round face whose bobbed black hair reaches just past her shoulders. Prior to her six years at Google she served in the powerful role of chief of staff to Lawrence Summers when he was secretary of the Treasury in the Clinton administration. She met him as a student at Harvard—yes, that school again—where she majored in economics. She wrote her thesis on the economic factors that lead women to remain in situations where they are abused by their husbands. (Zuckerberg has always sought to hire academic stars, despite having dropped out himself.) She speaks at an astonishingly rapid pace, yet without omitting inflection, in a kind of musical torrent of words. She's stylishly dressed when I come to see her, in new knee-high black Prada boots with black slacks and a cashmere sweater, and her polish contrasts dramatically with Zuckerberg's plainness—and with just about everybody else at Facebook.

Sandberg was eager to keep these meetings secret from others in the close-knit Silicon Valley community. One time she and her husband, Dave Goldberg, a top Yahoo employee, joined Zuckerberg and his girlfriend, Priscilla Chan, for dinner at an obscure restaurant near the San Francisco airport where no one would recognize them.

Zuckerberg asked a lot of questions, and Sandberg replied in kind. The topics they covered in their discussions ranged from where Facebook would be in five years to Sandberg's experiences in government to theories of management to personal history. He was vetting her, but she needed to be convinced as well. Shortly before they started meeting, the Harvard-oriented magazine 02138 had published a long exposé about the convoluted campus origins of Facebook. It accepted the Winkelvoss arguments at face value and suggested Zuckerberg was probably an intellectual thief. "We'll never know what really happened in the Harvard dorms four years ago," the article concluded. "The question remains: Whose idea was it?" Sandberg was concerned when she read it and queried her friend McNamee, who assured her of Zuckerberg's honesty.

Late in January both of them were heading to the World Economic Forum in Davos, Switzerland. Sandberg invited Zuckerberg to join her for the flight from San Francisco to Zurich on *Google One*, as the 767 owned by its co-founders Larry Page and Sergei Brin is known. The two talked conspiratorially the entire flight, a fact not unnoticed by some of her Google colleagues.

As the discussions got more serious, Sandberg called her good friend Don Graham at the Washington Post Co. to ask his opinion of Zuckerberg and Facebook. Graham had been among the many who tried to hire her in 2000 when she left the Treasury Department. (Others included the New York Times Company and the nonprofit AIDS Vaccine Initiative.) It turned out that Zuckerberg had also called Graham to ask about *her*. The Post CEO gave them both strong endorsements. "What a sensational hire that was. Wow," says Graham now about Sandberg at Facebook.

Jim Breyer also spoke at length with Sandberg and with others who were in contention for the COO job. She was one of the few who didn't say, one way or another, that she would like to keep open the possibility of being Facebook's CEO some day. That was a deal breaker. "Mark's our long-term CEO," says board member Breyer. "We were looking for a great business partner who was comfortable with that."

Aside from her willingness to be number two, the Harvard connection, the Graham connection, her role developing Google's ad business, and her experience as a manager, there was an additional thing that Zuckerberg found intriguing about Sandberg. "We spent a lot of time talking about her experience in government," he says. "In a lot of ways Facebook is more like a government than a traditional company. We have this large community of people, and more than other technology companies we're really setting policies." Beacon, of course, was an example of very poor policy setting.

He hired her, and she started at Facebook at the end of March 2008. If Microsoft's investment proclaimed to the world that Facebook was a formidable economic force, hiring this Internet superstar declared it would be a well-managed one.

For all the vetting and planning, on the day that Sandberg arrived she had some trepidation. What would it really be like working for this twenty-three-year-old day in and day out? On that first day, Sandberg, Zuckerberg, and the so-called M team—the group of eight or so most senior executives—were discussing a ratings system that was going to be used in human resources. The question arose: How does one best set up a ratings system? Sandberg had overseen many ratings systems at Google, so she spoke up. "You always have five categories: two on the top, two on the bottom, and one in the middle," she said briskly. Someone asked her why. "Well, three is too few, seven is too many, and six is an even number. You need one in the middle to anchor it. Everyone understands five categories," she asserted. Shortly thereafter the meeting ended. Zuckerberg walked out alongside Sandberg.

"I'm really sorry," he said.

"For what?"

"Well, I rolled my eyes."

"I didn't even notice."

"Well," Zuckerberg said, "I'm bringing you in here and I know I need to empower you and make sure everyone knows I believe in you, and I shouldn't be rolling my eyes."

She was impressed Zuckerberg would call himself out for such a minor infraction. "I said to myself, 'This is going to work,'" she recalls. And the candid back-and-forth between them has continued. They meet privately several times a week. For the first few minutes of every Friday's meeting they give each other direct feedback. Before Sandberg started she told Zuckerberg she wanted regular feedback from him. But he insisted it should go both ways.

From the moment she arrived, Sandberg was the company's top advertising champion and salesperson. She had immense experience with advertisers from Google and a deep appreciation of the importance and potential of ads on the Net. According to some at Facebook, in her first weeks there was hardly anyone else at the company about whom the same could be said. Despite the Microsoft deal, despite Facebook Ads, despite the clear need to build up revenue as the service burgeoned, there remained a profound corporate ambivalence toward advertising as the means for Facebook to become a real business. That ambivalence was rooted in the CEO, who firmly believed that the product and the user experience comes first. Sandberg had some work to do.

13

Making Money   "What business are we in?"

How would Facebook turn its social success into a lasting, moneymaking business? It was a question that could elicit a surprisingly broad range of answers even among senior executives at Facebook when Sheryl Sandberg arrived. Zuckerberg didn't have a good answer, though that didn't bother him much. But Sandberg, who is a very methodical manager, was intent on creating alignment among Facebook's leadership. She had come to the company to turn it into an advertising powerhouse. She needed all her staff and peers on the M team to work in synch. There was no question in her mind that Facebook represented one of the great advertising environments of all time.

The matter was hardly academic, because Facebook needed the money. It was burning through the $375 million it had raised from Microsoft, Li Ka-shing, and the Samwer brothers faster than anybody had expected. Some of Zuckerberg's allies in management had already concluded it had been an error not to accept a lower valuation, which would have allowed Facebook to raise a lot more money because so many more investors would have been willing to buy. The company had been hiring quickly and was by now paying about five hundred employees and adding servers to its data centers by the hundreds. Soon Facebook would also have to build new data centers outside the United States to accommodate its international growth. It had built a fancy new cafeteria for employees in a separate building a block or so from its main buildings, with chefs hired from Google and fabulous food—all served free. Plans were afoot to move out of the twelve buildings in which staff was now scattered throughout downtown Palo Alto and move into one big new space.

After Sandberg had been at the company about five weeks, she decided to host a series of meetings to get Facebook's management to focus on the ad opportunity. Zuckerberg wasn't going to be around, because he was embarking on a monthlong around-the-world trip, now that he'd completed his search for a number two. He'd wanted to take a break for a while. Now was his chance. He traveled alone, carrying only a backpack, to Berlin, Istanbul, India, and Japan, among other places. In India he made a brief pilgrimage—by dusty local bus—to the ashram high in the Himalayas where Steve Jobs and Baba Ram Dass, among others, have sought enlightenment.

Colleagues believe Zuckerberg timed the trip deliberately, to give Sandberg a bit of runway to establish her authority inside the company without his interference. But it is symbolically apt that her meetings about how Facebook could best turn its vast user base into a powerful business occurred with Zuckerberg—he of the ambivalence towards ads—out of town. Never before had executives from across the organization come together to brainstorm on what people in the Internet business peculiarly call "monetization"—how to turn all those Facebook users into money.

The meetings ran from 6 P.M. to 9 P.M., with dinner brought in, once or twice a week. The first one included a small number of the top ad-related leaders of the company: Mike Murphy, who headed ad sales; Chamath Palihapitiya, who was in charge of growth and international; Tim Kendall, who oversaw the online self-service ad business; Dan Rose, who managed the Microsoft ad partnership; Kent Schoen, head of advertising products; Kang-Xing Jin, the engineer responsible for advertising software (and Zuckerberg's close friend since Harvard days); and Matt Cohler, Zuckerberg's tousle-haired "consigliere." On the whiteboard Sandberg wrote, in big letters, "What business are we in?"

These were bull sessions at first, giving everyone a chance to express their views. As they continued, the meetings grew steadily in size. Word spread that you shouldn't miss these conversations. Pretty soon the entire M team and a larger swath of ad people were making it, a total of fifteen to twenty on a typical evening.

At the time, Facebook's monetization strategies were varied. Microsoft was selling banner ads, of course, but by the end of 2007, despite the new international deal, Microsoft accounted for less than 25 percent of overall revenue. Facebook wanted that figure down even further, so that it could control its own destiny. The self-service online ads launched at the same time as the disastrous Beacon were now growing rapidly. Facebook also had what it called "sponsored stories"—ads inserted into users' News Feeds that looked like an alert you'd get from a friend, except that it was from Coca-Cola or another company. Virtual gifts, a fast-growing but still tiny share of revenue, were little graphic icons people paid for. For your friend's birthday, for example, you could buy a little picture of a cupcake with a candle for a dollar. And finally there was the Facebook Marketplace, a classified ads system, which had only recently debuted to a lukewarm response from users.

On the whiteboard Sandberg listed the options. Facebook could be in the advertising business. It could sell data about its users. It could sell avatars and other virtual goods to those users. Or it could enable transactions and take a small cut, like PayPal. Staffers researched various markets and brought carefully compiled charts to the next meeting, showing the size of each market, its likely growth rate, the big players, and what Facebook could do uniquely well. After weeks of this, at the final meeting Sandberg went deliberately around the room and asked each person what percentage of Facebook's revenue would ultimately come from each category. Virtually everyone said 70 percent or more would be advertising in some form.

They all knew Zuckerberg only approved projects that fit into his long-range plan for Facebook. "Mark is very focused on the long run," says one participant in the meetings. "He doesn't want to waste resources on anything unless it contributes to the long run. If you don't know what business you're in, then anything you do to make money is a waste, because it might not last." While Zuckerberg had been forced by circumstances to accept advertising, he did so only so he could pay the bills. Whenever anyone asked about his priorities, he was unequivocal—growth and continued improvement in the customer experience were more important than monetization. Long-term financial success depended on continued growth, he believed, and even his grand declarations at the Facebook Ads launch just meant the company would start seeking new approaches. And the Beacon fiasco had shaken everyone's confidence.

In order to articulate a business strategy that would fit solidly and inarguably into Zuckerberg's long-term frame, the conferees at the Sandberg sessions went further than merely saying ads were it. They arrived at a crucial distinction to clarify and differentiate Facebook's opportunity. Whereas Google — Sandberg's corporate alma mater and the undisputed king of Internet advertising up to now — helped people find the things they had already decided they wanted to buy, Facebook would help them decide what they wanted. When you search on something in Google, it presents you an ad that is a response to the words you typed into the search box. Very often it's relevant to you and that process makes many billions of dollars for Google. But the ads you typically click on there are the ones that respond to what you already know you're looking for. In advertising-speak, Google's AdWords search advertising "fulfills demand."

Facebook's, by contrast, would generate demand, the group concluded. That's what the brand advertising that has long dominated television does, and that's where most ad dollars are spent. A brand ad is intended to implant a new idea into your brain — hey, you should want to spend money on this thing. But such ads have never worked well on Google. You may find a Canon camera via a Google search ad if you type the keywords "digital camera" in the search field, but the company has never found a good way to convince you that you should want a digital camera. (Google's efforts to find such methods are what have led it to emphasize, for example, its Gmail service, in which its software watches words in your emails and displays messages it thinks you might respond to.)

For all Google's success, it operates almost entirely within a relatively small sector of the overall advertising industry. Only 20 percent — at most — of the world's $600 billion in annual advertising spending is spent on ads aimed at people who already know what they want, Sandberg's researchers discovered. The remaining 80 percent, or $480 billion a year, was up for grabs as more and more ad spending shifted to the Internet.

The long-term prospects for advertising on Facebook looked bright to this group. The Internet is pulling consumers away from TV, newspapers, and magazines. And Facebook is taking a disproportionate amount of that Internet time. It is now where Net users spend the most online time by far in the United States and most other countries. That, says advertising marketer Dan Rose, combined with Facebook's unparalleled ability to target ads based on information about its users, should enable it to attract more and more demand-generation advertising as time goes on. Says Rose: "There is an imbalance between where the dollars are spent and where the audience is spending its time. Those dollars are going to move online over the next ten years." Rose was so effective in the sessions that afterward Sandberg gave him the new title of vice president for business development and monetization.

Sandberg's eight or so business-model sessions concluded just as Zuckerberg was returning from his round-the-world vacation. He was impressed with the group's conclusions. "Now Mark understands that we have a business model and this is the long-run thing," says one top ad executive. "So now he's willing to invest."

Zuckerberg explained his ideas about ads on Facebook in detail. At a subsequent off-site meeting on monetization, he told the group what made Facebook different from other websites was its ability to help users have two-way dialogues with one another or with advertisers. "The basic idea is that ads should be content," he says now. "They need to be essentially just organic information that people are producing on the site. A lot of the information people produce is inherently commercial. And if you look at someone's profile, almost all the fields that define them are in some way commercial—music, movies, books, products, games. It's a part of our identity as people that we like something, but it also has commercial value."

From these discussions with Zuckerberg emerged something Facebook calls the "engagement ad." It is a modest-looking message from an advertiser on users' home pages that invites them to do something right on the page. It might ask you to comment on a video in hopes that friends will be drawn into the conversation. It might be a product giveaway. Starbucks has offered coupons for free cups of coffee. It might enable you to engage in a dialogue with friends right there on the ad. Or it might enable you to click on the ad to instantly become a fan of a product's Facebook page.

Soon engagement ads replaced the sponsored story as the main product sold by Facebook's advertising salespeople. Sponsored stories are not, in Zuckerberg's terms, "organic information that people are producing on the site." The new engagement ads became a big hit. In the first year alone they generated close to a hundred million dollars of revenue. Facebook charges at least $5 per thousand views for these ads. With 400 million users viewing their home page many times a month, those dollars can add up. Moreover, once an advertiser establishes some sort of connection with a user it gets a tremendous amount of what Facebook calls "derivative value." Executives say that once a brand makes a connection with a consumer that leads to an average of about 200 free additional "impressions"—occasions when people on Facebook see information about that brand.

"We will never again sell banner ads," says Rose. "Engagement ads leverage the power of the Internet to enable the marketer to have a dialogue with the audience. That's very different from traditional banner ads on the Web. Those do what advertisers have done on TV and in print for fifty years—intentionally disrupt the experience you are having." Meanwhile, Microsoft continued to sell such banners for Facebook. They generated about $50 million in 2009, but the deal ended in early 2010, with Microsoft getting more involved in Facebook search in exchange.

But while the company has put great energy into developing and refining engagement ads and carefully managing its relationship with Microsoft, the lion's share of its ad revenue is coming from a third source: self-service ads that smaller advertisers purchase right on Facebook's site, using a credit card. Anyone can buy them, but these ads are typically purchased by local businesses.

Facebook gives advertisers more targeting options than most websites because people overtly and willingly put there a tremendous amount of information about themselves. They also spend a lot of time and do a wide variety of activities there, which creates opportunities to present people with ads. If on Google you buy an ad that displays when someone types "digital cameras," on Facebook you display a similar ad to married men in California who have young children but who haven't posted any photographs.

For all the importance of advertising, even in the Sandberg sessions there was another category of revenue that many believed will become quite large over time. Rose calls it "consumer monetization," meaning users pay Facebook directly for something, just as they already pay lots of money to play various games and other applications inside the service. Facebook was already selling things like virtual birthday cakes for a dollar, but there are many other ways Facebook could get money from its users. For example, there might be fees associated with a currency that people could use to buy and sell things across Facebook, especially in games. The company is testing such a system already. On other social networks around the world there is a healthy market in virtual decorations and virtual statements between friends. Sandberg says she believes that ultimately 20–30 percent of Facebook's revenues will come through the sale of virtual goods or the operation of an on-site currency. Virtual goods sales totaled an estimated $30 million in 2009.

In early 2010 Facebook began putting renewed emphasis on "Facebook credits," which users purchase from the company and then use mostly in games, to buy virtual goods. When a user spends the credit, Facebook keeps 30 percent of its value. Some games have begun using only this form of payment, replacing a plethora of third-party payment options that prevailed before. Justin Smith, who runs Inside Network, the leading analysis firm for Facebook commerce, says he believes such credits will become a major part of Facebook's future. "The idea," he says, "is that Facebook will be able to enable new kinds of revenue growth because users will be more comfortable paying Facebook than a third party." Smith even believes it possible that the company could eventually allow users to use Facebook credits for purchases across the Internet. Even a small cut of such a system might become a significant source of revenue. Zuckerberg, however, says the company work on credits thus far has been primarily to make life easier for application developers on Facebook's platform. "Our intention is not to be profiting off of this anytime soon," he says. "Over time, if it becomes a widely used thing, it could be a good business."

Facebook now sits squarely at the center of a fundamental realignment of capitalism. Mark Zuckerberg, as a man of his generation, has understood this intuitively since he launched Facebook at Harvard. Marketing cannot be about companies shoving advertising in people's faces, not because it's wrong but because it doesn't work anymore. The word *advertising* is no longer really the right word for what's going on at Facebook. It is merely a useful shorthand, as in the Sandberg sessions, to refer to a process in which companies spend money to get people more interested in their products.

But marketers can no longer control the conversation. It first became evident that consumers were becoming publishers when blogs emerged in 2001 and 2002. The audience was starting to create the media. Now Facebook is enabling that trend to broaden to even the least tech-savvy "consumer." Users get their own home pages and the tools to send messages and create and forward content. Much of that content is about commercial products and services. Anyone can also now create a Facebook page for any purpose.

Consumer spending is the engine that drives every modern economy. But "the consumer" no longer just consumes, as Facebook makes evident. Increasingly the people are in control.

"Brands are already on Facebook whether they like it or not," says Tom Bedecarre, who heads San Francisco's AKQA, the largest independent digital advertising agency, and an ardent fan of Facebook. "Whatever people hate or love they will start groups or pages about, and post messages about." One marketing tool often used by Facebook ad sales boss Mike Murphy is to search Facebook's database when he's trying to sell ads to a company, so he can demonstrate how enmeshed it already is on Facebook. For a well-known company like McDonald's, for instance, the number of mentions is in the millions.

Some companies make ill-fated attempts to squash consumer sentiment. Canadian coffee-shop chain Tim Hortons responded to Facebook groups that criticized the company by having lawyers send members cease-and-desist letters. That had little effect. No lawyer can prevent someone on Facebook from criticizing or insulting a brand or product. As Randall Rothenberg, president of the trade group Interactive Advertising Bureau, puts it, "Conversations cannot be controlled. They can only be joined."

Rather than interrupting the conversation, the companies formerly known as advertisers now have to figure out how to create the conversation on Facebook, or to be part of it. Successful ones help users connect to each other and communicate. "It's a new kind of exchange of value for marketers," says Bedecarre. "I'll give you value and you'll have a better feeling."

Mazda asked fans of its Facebook page to help it design a car for 2018. Design students from all over the world contributed ideas. Ben & Jerry's ice cream let people tell the company what its next flavor ought to be. Each time those Mazda or Ben & Jerry's fans write something on those pages, a message is posted on their profile that goes into friends' News Feeds. Consumers are sending messages to their friends that benefit the marketer. That's how the flavor program, developed by marketing firm Edelman Digital, enabled Ben & Jerry's to increase its fans from 300,000 to a million in just six weeks. The campaign in both cases began with engagement ads on Facebook's home page.

Facebook users often get something concrete as they're being marketed to. In effect they are receiving some of the compensation that would otherwise have gone to a TV station or newspaper in the past. Starbucks has given away coupons for free cups of coffee. Ben & Jerry's has given away ice-cream cones. Giveaways have worked for marketers seeking to reach business customers, as well. AKQA helped client Visa create the Visa Business Network for small businesses on Facebook. Visa gave each company that signed up $100 worth of Facebook advertising. Several hundred thousand did.

Some consumer-oriented companies now put less emphasis on their website and more on their Facebook page, where they can host a wide variety of Facebook applications and where actions of fans get virally projected to their friends. Vitamin Water, for example, has begun to direct consumers to Facebook.com/vitaminwater from its TV ads and from banners placed elsewhere around the Web. Gap displays the address of its Facebook page on billboards.

The relationship between people and companies will continue to evolve rapidly on Facebook, and will most likely yield some startling developments. There's growing evidence that by enlisting consumers into the very process of conceiving, designing, and even building a product, companies can reduce their costs, create products people want, and engender customer loyalty. Facebook can be seen as a giant collaborative network. It is the perfect platform for such innovation. The competitions from Ben & Jerry's and Mazda pointed the way, but in 2009 a small film company called Mass Animation, working closely with Facebook staffers, took the idea considerably further.

It produced an animated film created by the users of Facebook. The five-minute film, titled *Live Music,* includes segments contributed by fifty-one different people from seventeen countries, including Kazakhstan and Colombia. Some were as young as fourteen. Mass Animation created a storyline, soundtrack, and first scene, which established the film's graphic style. Its Facebook page attracted 57,000 members, 17,000 of whom downloaded special software. Members of the page voted to determine which segments should be included in the film. Winning contributors received $500 and acknowledgment in the film, which Sony distributed to theaters in late 2009 as the opener for an animated feature. "Social networking is becoming social production," says Don Tapscott, an author who wrote both *Wikinomics,* about new forms of business collaboration, and *Grown Up Digital,* about young people and technology. "This is not just about friendships. This is changing the way we orchestrate capabilities in society to innovate, and to create goods and services."

Facebook is the most targetable medium in history. Advertisers want to show their ads to the people who are most likely to respond. On the Net, until Facebook came along they had to hire services to laboriously and expensively follow users' digital footprints across the

Internet, attempting to infer their gender, age, and interests by where they visited and what they clicked on. But on Facebook users are forthcoming with accurate data about themselves, because they are confident the only people who will look at it are those they approve as their friends. "Facebook has the richest data set by a mile," says Josh James, CEO of Omniture, a big Internet ad-targeting service that works with Facebook. "It is the first place where consumers have ever said, 'Here's who I am and it's okay for you to use it.'" Sandberg says, "We have better information than anyone else. We know gender, age, location, and it's real data as opposed to the stuff other people infer." The inferential targeting used by advertisers on the rest of the Web is frequently wrong, she says.

Users on Facebook do volunteer vast amounts of data about themselves, and then generate even more through their behavior on the site, by interacting with other users, on groups and with pages. Facebook tracks all this in its database and uses it to place advertisements. Facebook's policy is not to look at any individual's data except to ensure it does not violate the service's rules. It says it never shares the actual data with advertisers. Facebook just lets advertisers use the aggregated data to select from a vast menu of parameters to target ads at precisely the type of person they are trying to reach.

Anybody can pick through endless combinations on Facebook's self-service ad page. You can show your ad only to married women aged thirty-five and up who live in northern Ohio. Or display an ad only to employees of one company in a certain city on a certain day. (Employers aiming to cherry-pick people from a competitor do this all the time.) Customers for Facebook's more expensive engagement ads can select from even more detailed choices—women who are parents, talk about diapers, listen to Coldplay and live in cities, for example. "That targeting pure and simple is the driver of what we're able to do today, and why we're growing," says Facebook's Rose.

I am a baby boomer and list many musicians I like on my profile. So I frequently see an ad on Facebook for a USB turntable that converts old vinyl records to digital MP3 files. The advertiser targets music lovers my age because we're likely to own a lot of records.

The knowledge Facebook has about its users enables it to help advertisers with market research. Say a company is deciding what music to use in a TV ad. Facebook can survey the profiles of all the people who are fans of that advertiser's page and report what music they are most likely to listen to. If you buy an engagement ad, Facebook can tell you the exact demographic breakdown of the users who clicked on it. "I can tell an advertiser, for example, that while it thought its audience was eighteen-to-twenty-four-year-old women, they are actually nineteen-to-thirty-eight-year-old men," says Facebook ad boss Mike Murphy, "and they like football and these are their three favorite movies. If you want to reach these guys, here are their favorite TV shows. You can build your entire media campaign around the data we provide you. It's an asset you couldn't buy anywhere else on the planet." Now the company is working with a service called Nielsen Homescan to correlate data Nielsen collects about product purchases in thousands of American homes with the Facebook behavior of those residents. Advertisers will be able to see which ads Facebook users saw and which products they bought. That sort of data has existed for a long time for television. If Facebook can demonstrate it is at least as effective, advertisers will become more eager to be there.

Facebook's ability to marshal all that user-reported data makes some believe it can make a lot of money. "Facebook has the opportunity that Google only wishes it had—the ability to build a credible proposition for the largest brand advertisers," says Alan Gould, who runs ad-measurement firm Nielsen IAG. "Now Steve Ballmer's valuation doesn't look so silly." "I believe Facebook is going to fundamentally change marketing and become a monster business," says Mike Lazerow, CEO of Buddy Media, which builds promotional Facebook applications and pages for companies. "When you combine four hundred million people with data about not only where they live, but who their friends are, what they're interested in, and what they do online—Facebook potentially has the Internet genome project."

So far there has been little resistance among Facebook's users to using their data to target ads to them. But it could be where the privacy challenge becomes greatest. It's easy to imagine how some error of targeting or other clumsiness could lead to a major ad backlash that sullies the company's reputation.

Not that there haven't been problems. In this world of marketing centered on the likes and dislikes of actual people, the biggest danger so far has been that users would appear to endorse or to initiate the transmission of messages that they actually disapprove. One man named Peter Smith from Lynchburg, Virginia, noticed in July 2009 a Facebook ad reading "Hey Peter—Hot singles are waiting for you!!" Next to it was a photo of an attractive, smiling woman—who happened to be his wife. It turned out Cheryl Smith played games on Facebook. She had given a game permission to access her data, through the opaque process Facebook uses to connect users to applications. The game company used a third-party network, which displayed ads inside the game.

Apparently the ad network appropriated her picture from inside the game and affixed it to the dating ad. The ad network that stole the picture was violating Facebook's rules and was banned. Facebook subsequently clarified its advertising guidelines to make clear that such sharing of user data is not allowed. But as people interact with applications and use Facebook in a larger variety of ways it has become increasingly harder for the company to police how user data is handled. More mistakes are bound to happen.

In the months after Sandberg arrived at Facebook, the company's leadership went through a fundamental realignment. There was a string of departures. Owen Van Natta was the first to go, not surprisingly. It was obvious that no matter what happened, with Sandberg's arrival he wasn't going to get a shot at CEO. Within a year Van Natta became CEO of MySpace (though he lasted there less than a year).

As Sandberg settled in and refocused Facebook on its fundamental opportunity in advertising, Zuckerberg's founding team—the young posse who had helped him create Facebook—also began to disperse. Matt Cohler, his "consigliere" since early 2005, left to join prestigious Benchmark Capital and become a venture capitalist, something he says he'd always wanted to do. He remains close to Zuckerberg. Adam

D'Angelo, Zuckerberg's Exeter chum who has come and gone from Facebook several times, left again to start a new company called Quora, and took top engineer Charlie Cheever with him.

But most striking was the departure of Dustin Moskovitz, Zuckerberg's right-hand man since the very beginning, and still one of the company's largest shareholders, with about 6 percent of the stock. Moskovitz, like D'Angelo, remains close to Zuckerberg. Moskovitz left to start his own Internet software company called Asana, an idea he'd been mulling for a long time. He aims to build Facebook-connected online productivity software for businesses, competing with Google Docs and Microsoft Office, among others. It's a big and ambitious vision. He says he thought for a long time about whether he could stay at Facebook while pursuing this new idea, but concluded it would be a distraction for the company.

The influence Moskovitz wielded as the self-taught roommate-turned-CTO inevitably waned as the company passed one thousand employees and everything became more professional. There was a long time when he jointly controlled the direction of the company. But as it grew, Zuckerberg's authority grew along with it, and Moskovitz's diminished. Despite his large stockholdings he cannot have the impact he once did. "There are just disagreements about the direction the company goes in," says one friend of both men, "and when you've got someone who has sole authority, those disagreements are irreconcilable." It also made sense for Moskovitz to start Asana outside the company because Zuckerberg has repeatedly shown he has little interest in adding features that make Facebook more useful in the workplace.

In each case, Zuckerberg's close friends—and they all still call themselves that—say they didn't leave because of any fundamental conflict with Mark. D'Angelo says he is just not suited for large organizations where compromise is constantly required. He says he remains very attached to Facebook but got frustrated with the bureaucracy he had to deal with every day. Zuckerberg "just has a lot more tolerance for that than someone who doesn't feel like it's their company," D'Angelo says.

Chris Hughes, the other co-founder, who had left the company earlier, is more blunt. He thinks Zuckerberg's friends, most of whom he's in touch with, have left in part because, like him, they got fed up. "Working with Mark is very challenging," says Hughes. "You're never sure if what you're doing is something he likes or he doesn't like. It's so much better to be friends with Mark than to work with him."

The CEO is a tad melancholy about the departure of his boys. He says he was upset when Moskovitz first told him he wanted to go, a year before he actually left. By the time it happened Zuckerberg was resigned to it. As for Cohler and D'Angelo, Zuckerberg says, "I wish we'd been able to figure out a way to continue finding them roles."

Bringing in Sandberg as number two had little to do with this posse heading for the hills, but Moskovitz, for one, hasn't signed on to the enthusiastic consensus that emerged from the Sandberg sessions. He responds with typical directness when I ask him about Sandberg's impact on Facebook. "Positive overall for sure," he starts out. Then he continues, equivocating. "It's hard for me to be too positive, because I do feel like her role is in conflict with what I think the natural course of the company is. At the same time, I very much understand. But I am a huge believer in investing as much as possible into the product, do as little as possible to provide friction against more people joining or not liking the experience as much. And that can often be in direct conflict with the amount of advertising on the page, which is her job responsibility." He says it's good that she clarified how ads would work on Facebook, but adds, "I see that as just like a necessary evil, almost." Then he backs off a bit and concedes, "It's probably the right balance now."

Despite the consensus that Facebook's business is advertising, Zuckerberg continues regularly to declare that growing Facebook's user base remains more important than monetizing it. And both Moskovitz and D'Angelo continue adamantly to agree with him. "You can make a dollar off a user today," says Moskovitz, "but if you can get them to invite ten friends, then you'll make eleven dollars. Facebook's growth is so exponential that it's really hard to say this is the point at which you start compromising." D'Angelo also shows little enthusiasm for emphasizing ads now. "I'm on the growth side, personally," he says. "I mean, if you think Facebook is going to be around for a long time, which I do, and you take this approach that we need to get this thing to be everywhere and get the whole world using it, then to me it's obvious you will make a lot of money off of a product that the whole world is using every day."

Zuckerberg's top anti-ad, pro-growth allies have retreated, but he remains deeply committed to the long-term view. "It's really important for people to understand that what we're doing now is just the beginning," he says. "The companies that succeed and have the best impact and are able to outcompete everyone else are the ones that have the longest time horizon." Board member Peter Thiel has always been another strong believer in the need to continually emphasize growth. Even at some points in the company's history when Zuckerberg was focusing on other matters, Thiel repeated his steady refrain: "Grow the user base. Grow the user base."

Mike Murphy, Facebook veteran and hard-charging sales guy, concedes there has been ongoing tension over whether revenue mattered as much as growth, and that it drove him crazy after he arrived in early 2006. "My level of frustration has decreased dramatically," he says now. "Mark has never missed a commitment he's made about resources he would give us." The company has about 260 people devoted exclusively to ad sales. Before Sandberg arrived, Facebook only had sales offices in Palo Alto, New York, and London, but in the year following the Sandberg sessions it opened offices in Atlanta, Detroit, Chicago, Dallas, Dublin, Los Angeles, Madrid, Milan, Paris, Sydney, Stockholm, Toronto, and Washington. Shortly the company plans to add more in Boston, Germany, Hong Kong, India, and Japan. Its international headquarters is in Dublin.

Sandberg says that a focus on growth does not conflict with a mandate to raise revenues. "Our goals are, in order: How much does the world share information? Then, of equal importance, How many users do we have? And revenue. Those are all really really important drivers of the whole mission. But you can't do one without the other."

The ad industry is shifting its focus toward Facebook. The number of advertisers using its self-service online ads tripled from 2008 to 2009. A 2009 study by the Association of National Advertisers found that 66 percent of all marketers now use social media in some way, compared to only 20 percent in 2007. Today that mostly means Facebook. The vast majority of the biggest advertisers in the United States have begun advertising there. Big clients include PepsiCo, Procter & Gamble, Sears, and Unilever. And Facebook users are embracing the growing commercial presence on the site. Pages had about 5.3 billion fans as of February 2010 and about twenty million users become new fans of Pages every day. Pages with more than 3 million fans include Coca-Cola, Disney, Nutella, Skittles, Starbucks, and YouTube.

The mood inside the company about Facebook's financial prospects is bright. Marc Andreessen, whom Zuckerberg asked to join the company's board of directors in early 2008 (to fill one of the empty seats), cannot say enough about how big Facebook's business can be. "Facebook has a springboard to monetization that is as clear as anything I've ever seen," he says. "Like night follows day. With TV, radio, magazines, and newspaper revenues dropping, there's $200 billion of ad spending up for grabs. That money has to go online. And Facebook's just going to have all this data as a consequence of all the user activity, and it's going to be able to target against that." Television became the recipient of the lion't share of ad dollars because that's where consumer attention was focused. If that attention is slowly shifting to a new medium, as the data suggests, so will the money.

Sandberg was surprised that Facebook's business did so well during the recent economic downturn. In the fall of 2008 the company significantly reduced its goals for growth and cut planned spending. "The world looked like it was melting down, and I was nervous," says Sandberg. It seemed inevitable the global recession would hurt Facebook. It didn't. In an interview in mid-2009, she said, "Our ad rates are basically holding, in an era when everyone else is dramatically decreasing theirs. We're just doing better and better and better." The measurement firm comScore reports that U.S. online advertising is moving to social networks—they now garner 23 percent of total ads—and that Facebook displayed 53 billion ads in December 2009, or 14 percent of *all* online ads.

Sandberg's efforts to bring clarity to Facebook's business model are paying off. She has found her place in this youthful culture. Other top managers both on and off the record express admiration for how well she runs the organization, interacts with people, and gets things done.

Now Facebook's numbers are rising rapidly. While Facebook does not disclose its financials, overall revenues were, according to well-informed sources, more than $550 million for 2009—up from less than $300 million in 2008. That represents a stunning growth rate of almost 100 percent. The same sources say that the company could exceed $1 billion in revenue in 2010.

Facebook's improving numbers are fueled especially by its highly targeted online self-service ads, sold mostly to smaller advertisers, for all the efforts devoted to larger advertisers are still the lion's share of revenue. Between $300 million and $400 million came from those in 2009. While the prices Facebook can charge for such ads remain very low on average, the company displays so many of them that it is becoming an increasingly good business. Says one well-informed company insider: "People dramatically underestimate the impact on our revenue of two interrelated factors—the growth in the number of users and the growth in usage." Research firm comScore calculated in late 2009 that the average Facebook user in the U.S.—and there are almost 110 million of them—spends six hours per month on the service.

The next largest category is engagement ads and other brand advertising sold directly by Facebook, which probably amounted to about $100 million. Ads sold by Microsoft represent another chunk—more than $50 million. Finally, virtual goods and other miscellaneous revenue accounts for between $30 million and $50 million.

"There has been this myth that everyone's waiting for our revenue model," says Sandberg. "But we have the revenue model. The revenue model is advertising. This is the business we're in, and it's working." Few at Facebook disagree with her now.

14

Facebook and the World "Making the world more open is not an overnight thing."

Mark Zuckerberg is in a large van on the campus of the prestigious University of Navarra in Pamplona, Spain. It's October 2008 and he's just finished speaking for an hour in the school's largest lecture hall. The hall seats only about four hundred, but at least six hundred students had crammed inside. Before the van can move, a crowd gathers, all of them waving frantically and straining to catch Zuckerberg's eye. As the van pulls away, a group of five or six girls runs ahead. When he gets out at his next destination, the president's office—the girls are there again. Zuckerberg amenably agrees to pose with them for a photo (to be posted on Facebook, of course). Then the group dissolves into elated giggles, still casting sidelong glances, not believing their good fortune. "You're a rock star now," says Anikka Fragodt, Zuckerberg's trusted personal assistant (since February 2006), who with three other Facebook employees (and me) has joined him for a promotional swing through Europe.

An epochal change on the Internet was announced in March 2009 by the Nielsen Company research firm. Time spent on social networks by Internet users worldwide had for the first time exceeded the amount of time Internet users spent on email. A new form of communication had gone mainstream. Total time spent on social networks grew a healthy 63 percent in 2008 around the world. Facebook, however, was in another league. It outdistanced every other service Nielsen measured. Time spent on Facebook had increased 566 percent in a year, to 20.5 billion minutes.

The scale of Facebook's global growth in recent years is difficult to grasp. From the moment it opened to nonstudent users in fall 2006, English-speakers around the world began to stream on board. In early 2008, Facebook inaugurated a novel translation project, and by the end of 2008 it could be used in thirty-five languages. But even then, with the internationalization project still in its early phases, 70 percent of Facebook's then 145 million active users were already outside the United States. Nielsen calculated at that point that fully 30 percent of the world's Internet users were on Facebook, up from 11.1 percent a year earlier. The only service with more users is Google.

The company's own expectations continue to be surpassed. Its ambitious confidential internal goal at the beginning of 2009 was to reach 275 million active users by the end of that year. Few at the company thought it attainable. But it reached the goal by August and by the end of the year had more than 350 million users and was growing about a million new users per day in 180 countries.

Improbable statistics continue accumulating. In seventeen countries around the world, more than 30 percent of all citizens—not Internet users but *citizens*—are on Facebook, according to the *Facebook Global Monitor.* They include Norway (46 percent), Canada (42 percent), Hong Kong (40.5 percent), the United Kingdom (40 percent), Chile (35 percent), Israel (32.5 percent), Qatar (32 percent), and the Bahamas (30.5 percent). In tiny Iceland, 53 percent of people are on the service. Facebook is the number-one social network in Brunei, Cambodia, Malaysia, and Singapore, among other countries. It surpassed MySpace in global visitors in May 2008, according to comScore. And in mid-2008 the word *Facebook* passed *sex* in frequency as a search term on Google worldwide.

It's been a joke around the Facebook offices for years that the company seeks "total domination." But the reason it's funny is that it evokes a surprising truth. Zuckerberg realized a long time ago that most users are not going to take the time to create multiple profiles for themselves on multiple social networks. He also knew from his endless bull sessions at Harvard and in Palo Alto about "network effects" that once consolidation begins on a communications platform it can accelerate and become a winner-take-all market. People will join and use the communications tool that the largest number of other people already use. He therefore made it a goal to create a tool not for the United States but for the world. The objective was to overwhelm all other social networks wherever they are—to win their users and become the de facto standard. In his view it was either that or disappear.

Other social networks have more users than Facebook in a number of key countries, including Brazil, China, Japan, Korea, Russia, and a few other places. In most of those countries a local player commands the market. For Zuckerberg it is a strategic imperative to whittle away at the dominance of these services. As Zuckerberg told a Madrid audience on his Spanish trip, "Making the world more open is not an overnight thing. It's a ten-to-fifteen-year thing."

But how did Facebook get so big so fast? It wasn't long after he moved to California that Zuckerberg began thinking about Facebook's potential to be a global phenomenon. Influenced by the ambitious Sean Parker, Zuckerberg began to think that if he managed his service well it could grow into an international colossus. He did a lot of things right that set the groundwork for the vast global growth that followed. For one thing, Zuckerberg kept Facebook's interface simple, clean, and uncluttered. Like Google, an elementary look successfully masked an enormously complex set of technologies behind the curtain and made a wide variety of people feel welcome. At one of his stops in Spain, Zuckerberg summarized his international strategy: "It's just to build the best, simplest product that lets people share information as easily as they can."

Facebook also has a fundamental characteristic that has proven key to its appeal in country after country—you only see friends there. It is Facebook's identity-based nature that differentiated it from the beginning from most other social networks and enabled it to become a unique global phenomenon. Around the world this is the least American-feeling of American services. Italy's Facebook-using hordes, for example, could grow to many millions without often seeing anyone who wasn't Italian. The values, interests, tone, and behavior that users in Turkey or Chile or the Philippines experience inside Facebook are the same ones they are familiar with every day in the offline world.

And, critically, the language people speak on Facebook is increasingly the one they speak offline as well. The translation tool Facebook made available after early 2008 was among the company's greatest product innovations and had huge impact on its global growth. By early 2010 Facebook operated in seventy-five languages, representing 98 percent of the world's population.

Facebook's translation tool adopted a novel approach that took advantage of the rabid enthusiasm of users around the world. Rather than ask its own employees or contractors to spend precious years translating the site's three hundred thousand words and phrases into numerous other languages, Facebook turned the task over to the crowd and found an enormous amount of wisdom there.

To create a version in each new language, Facebook's software presents users with the list of words to be translated. Anyone, while using the site, can tackle the Spanish or German or Swahili or Tagalog translation for just one word or as many as they choose. Each word is translated by many users. Then the software asks speakers of that language to vote on the best word or phrase to fill each slot.

The tool was first used for Spanish in January 2008, since Facebook at that point already had 2.8 million users in Spanish-speaking countries using it in English. Within four weeks, 1,500 Spanish speakers from around the world had created a full version. Facebook engineers just plugged in their conclusions and the Spanish Facebook launched on February 11. Next up was German. That took 2,000 people two weeks and began operating on March 3. The French version was completed by 4,000 users in less than two days. Adding new languages now costs Facebook virtually nothing. Users decided the idiosyncratic Facebookism *poke* should become *dar un toque* in Spanish, *anklopfen* in German, and *envoyer un poke* in French.

This is one project Zuckerberg didn't oversee. "I'm proud that I wasn't even involved," he said around the time the translator launched. "This is what you hope for when you're building an organization, right? That there will be people who will just build things that fit so well with the values of the company without you even having to say anything."

Facebook's platform strategy of letting outsiders build whatever applications they want on its platform also substantially benefited its international expansion. In July 2008 the company let developers start using the translation software for Facebook applications, so those too could be available in any language. By the fall of 2008, when Zuckerberg went to Spain, there were already over six thousand applications available in Spanish. Facebook in Spain—or Chile or Colombia—felt much like a Spanish service to its users there. Eight months after the debut of the translated version, Facebook's Spanish-speaking population had more than quadrupled to 12 million. "We think we can get as much as thirty-to-forty percent of the population using it," Zuckerberg told reporters in Madrid. (Spain alone has 46 million people.)

There's almost a moral component to Zuckerberg's globalization quest. In the packed, sweltering hall in Navarra he says Facebook is "for all people of all ages around the world." Giving people more information about people around them "should create more empathy." In this attribution to Facebook of a power to help people better understand one another, Zuckerberg has a surprising ally—his mentor and board member Peter Thiel. The hedge fund manager and venture capitalist thinks Facebook is a key tool for a world necessarily becoming much smaller. "People in a globalized world are going to be in closer proximity to each other," he explains. "The key value in my mind will be more tolerance. What I like about the Facebook model is it's centered on real human beings and it enables them to become friends with other people and build relationships not only in the context they're already in, but in contexts outside of that as well. Globalization doesn't necessarily mean you are friends with everybody in the world. But it somehow means that you're open to a lot more people in a lot more contexts than you would have been before." At another session in Spain, Zuckerberg answered a reporter's question about why Facebook succeeded by saying, "If you give people a better way to share information it will change people's lives."

But Zuckerberg's Facebook is resolutely American, even if it may not always seem so to its international users. Facebook's Americanness is revealed not because some Azerbaijan teenager meets a kid from Oklahoma, but by its intrinsic assumptions about how people ought to behave. Zuckerberg's values reflect the liberties of American discourse. Facebook carries those values around the world, and that's having both positive and negative effects.

In the United States, people take a certain amount of transparency and freedom of speech for granted, but it comes at great cost in some other cultures. When a father in Saudi Arabia caught his daughter interacting with men on Facebook, he killed her. Users in the United Arab Emirates created protest groups with names like "Gulf Air Sucks," and "Boycott Dubai's Dolphinariums." That was apparently within the bounds, but when groups there grew to include "Lesbians in Dubai," with 138 members, the government attempted to ban Facebook altogether.

Governments around the world are struggling to figure out how to handle Facebook's users when they take advantage of its freedoms. After Italian Facebook groups emerged praising imprisoned mafia bosses, a senator there introduced a bill that would force websites to take down content that "incites or justifies" criminal behavior. It did not pass. (Facebook's own policies are more specific. It takes down content that advocates hate, violence, or breaking the law.)

In the West Bank, protesters directed their wrath at Facebook itself and drew it into delicate matters of international politics. Jewish settlers in the occupied territory were outraged that Facebook required them to say they lived in Palestine. A group called "It's not Palestine, it's Israel" quickly acquired 13,800 members in March 2008. After a few days Facebook agreed to let residents of certain large settlements say they lived in Israel. Meanwhile, a group called "All Palestinians on Facebook" grew to 8,800 by complaining, among other things, that Palestinians living in East Jerusalem were forced by Facebook to say they lived in Israel, even though that country's annexation of East Jerusalem has not been internationally accepted. Now Facebook users in the West Bank can say they live in either Israel or Palestine.

American values of transparency may not always translate well, but people in many cultures are embracing fuller disclosure about themselves. In the Philippines, it has become routine for middle-class people to post photos of their April and May summer vacations to Facebook, and to keep friends apprised about these trips with status updates. By late 2008, interacting on Facebook was so popular in Italy that Poste Italiane, the national postal service, started blocking access in its offices. (Employees of the city of Naples, however, were officially allowed to access Facebook for up to one hour per day.)

Cultural differences seem not to deter people in various countries from finding compelling uses for the service. Danish prime minister Anders Fogh Rasmussen had 12,000 supporters on his Facebook page in April 2008 and responded personally to every comment. Then he decided to set up a group jog with young people he met there. An aide called it a great way to connect with ordinary voters. Obscure Colombian rock bands like Koyi K Utho, which plays heavy-metal music inspired by Japanese anime cartoons, found an audience on Facebook to promote concerts and albums.

One aspect of Facebook's Americanness was an advantage, especially in its early years among students outside the United States. Its academic roots at Harvard and the Ivy League made it seem even more appealing. "I've heard people at Facebook say they worried that it made them seem elitist, but in fact many kids around the world put those schools up on a pedestal," says Jared Cohen, author of *Children of Jihad*, an account of how youth in the Middle East view culture and technology. As early as mid-2007, Facebook was being used by 20,000 English-speaking Egyptians, for example, mostly privileged, Western-oriented college students and recent graduates. "I log in three hours a day, more or less, and usually at night, too," Sherry El-Maayirgy, a Cairo marketing executive, told the English-language magazine *Egypt Today* in May 2007. "It is really an amazing place to meet new people and catch up with old friends who have drifted away." Much of the online behavior was libertine. A local group called "If this group reaches 1,000 members, my girlfriend will sleep with me" garnered supportive comments, according to the magazine. And beauty contests proliferated, such as one for "The hottest girl at the American University of Cairo."

Facebook's growth around the world belies the frequent American misperception that it is a site primarily for young people. While in the

United States many adults still spurn the service or quickly tire of it, in most other countries it's used by people of all ages. Facebook's greatest global increase in 2008 came from people ages thirty-five-to-forty-nine, according to Nielsen. That group now constitutes about a third of Facebook's users. "Internationally . . . Facebook is perceived as mainstream and MySpace as being more focused around a younger demographic," says the Nielsen Company in a report on global social networking. Facebook seems to mirror real-world conditions. Women account for more than half of Facebook's ranks all over the world—except in certain countries in the Middle East and Africa where their rights are severely curtailed.

In some countries Facebook's empowerment of the individual may feel even more important than elsewhere. Educated young people in the Middle East are often passionate and active Facebook users. "Kids there have some of the most intricate profiles," says Cohen. "These are repressive countries, with little outlet for expression, so people can feel more real online than they are in real life." Facebook can become a way to assert one's right to be oneself. In both Turkey and Chile, Facebook is so ubiquitous in many educated circles that not to be on it is tantamount to self-ostracism. One reason may be that in both countries not long ago, to oppose the government could lead you to disappear forever.

Facebook continues to face potent rivals. MySpace is really not one of them, having shifted its strategy under the leadership of Owen Van Natta, Zuckerberg's former chief operating officer. MySpace now emphasizes its role as a portal for music and entertainment. More worrisome are social networks that dominate in one country or region. In Japan, leading social network Mixi offers a sophisticated service that works as well on cell phones as on PCs. It specializes in games.

Orkut still leads by a large margin in Brazil. It also led for a long time in India, though Facebook surpassed it in popularity in late 2009, according to the Alexa Internet data service. Orkut's peculiar success in these two markets has led to a surprising new sort of Indian pilgrimage—young Indian men trek by plane to Brazil to see women they met on Orkut. In India, Facebook has now introduced versions not only in Hindi, the largest language, but also Bengali, Malayalam, Punjabi, Tamil, and Telugu.

Displacing Orkut in Brazil may turn out to be its ultimate popularity contest, but in the meantime Facebook faces tough battles elsewhere. In Germany, Spain, Russia, and China, local entrepreneurs created student-focused networks modeled explicitly after Facebook once its U.S. popularity became apparent in 2004 and 2005. While Facebook has now surpassed its clone rival Tuenti in Spain, domestic imitators in China, Germany, and Russia still command dramatically more users.

The hapless Friendster, essentially ignored in the United States, was until recently Facebook's big obstacle in Southeast Asia, where 90 percent of Friendster's 105 million users were located as of mid-2009. But by late 2009 Facebook had trounced it there and was the number-one website of any type in Indonesia, Malaysia, and the Philippines, Friendster's three biggest countries.

China's largest domestic Facebook clone, Xiaonei (the name means "in the school"), got a big boost in 2008 when Japan's Softbank Venture Capital invested $430 million in its parent company. It then renamed itself Renren, meaning "everyone," to broaden its appeal. Meanwhile, since June 4, 2009, the twentieth anniversary of the Tiananmen Square massacre, Facebook has been completely blocked in China by the government.

Part of Facebook's arsenal against Renren (and Friendster) is Facebook's close partnership with Hong Kong billionaire Li Ka-shing. Among the many companies this mogul controls is Hutchison Whampoa, a major provider of mobile telephone services across southern Asia. Hutchison has already released a special "Facebook phone" for the region. Social networks are used most commonly on mobile phones in countries like India and Indonesia, so Facebook is creating partnerships with local mobile operators. It has also released a so-called lite version that gives users the basics (without video, chat, and some other features) but requires little bandwidth. It can be used on mobile phones or where Internet access would otherwise be inadequate.

Facebook is just beginning to model itself to suit the preferences of users in one country. For example, in Germany, Facebook has a deal with the dominant local email provider to make it easier to register and connect with the friends in your email address book. In Japan the site will soon make it easier to blog and to operate on mobile phones. Executives are thinking of ways to accommodate the reluctance of Japanese to operate openly online using their real names, even though that will remain the way to use Facebook.

Facebook has exploded across Asia in the last year or so, but for different reasons in each country. In Indonesia, Friendster had been the dominant local social network, but as Internet usage shifted to mobile phones, Friendster didn't have a good mobile app. Facebook did, and burgeoned. In Taiwan, Facebook usage—mostly on PCs—soared in 2009 for one reason: the Farmville game from Zynga. Playing it became almost a national obsession, and many joined Facebook simply to do so. It grew from almost nothing to 5.6 million people, or 26 percent of the population in the year ended February 2010. In Malaysia Facebook took off among the influential Chinese minority, while those of Malay ancestry tended to stay on Friendster. As of February 2010 Facebook was growing 10 percent per month in Malaysia, according to the *Facebook Global Monitor*. What makes this growth more impressive is that it occurred without the investment spent by earlier American Internet companies, says Hong Kong–based social media expert Tom Crampton of Ogilvy Public Relations. "Facebook's romp across Asia is an amazing story that breaks all the rules of internationalization," he says. "When Yahoo entered Asia it sent huge teams to each country."

Scale itself is a growing advantage for Facebook. Sophisticated social networking features cost money to develop. But every line of software code on Facebook can be used by far more people than a comparable line of code on any other service. It is no longer possible as it once was for rivals simply to steal the Facebook software they want. So on a per-user basis Facebook costs less to run, and less to improve. That could prove over time to be a daunting advantage against its rivals.

The strength of regional competitors outside the United States is the biggest reason why Zuckerberg says that near-term growth is more important for Facebook than profit. He's not a worrier, but if he worries about one thing it's that nationalism and insular local cultures will allow services like Renren and Orkut to keep Facebook down. A couple of days before I joined him in Madrid, he gave an interview in Germany in which he said bluntly that "growth is primary, revenue is secondary." The statement was immediately criticized online as naive, and everywhere I went with Zuckerberg he was hounded about it by bloggers and press.

The only reason Zuckerberg is willing to endure the discomforts of a multiweek European road show is that he feels so passionately about the need for Facebook to grow internationally. He would prefer not to stand up and talk to crowds. But if that's what it takes, he'll do it. As he walks into a meeting in Madrid with a group of local entrepreneurs, his host welcomes him saying, "There is great expectation for your visit!" "That's unfortunate," Zuckerberg deadpans in a serious-sounding voice, as his staff cringes.

He's on the road with a purpose, but he does it in his own way, sometimes to his detriment. The trip wears on him. He was up doing email and instant messaging the previous night until four. Back in the van, his assistant Anikka Fragodt says he should take a nap. He doesn't think that would help. Anyway, he hates to remove his contact lenses. At the next stop, Madrid's University of Comillas, he is greeted by two deans. One of them holds out a soccer jersey with the university's logo on it. Zuckerberg refuses to put it on. "This is what I always wear," he says of his black North Face fleece jacket, T-shirt, jeans, and running shoes. At Navarra a few days later, the lecture hall gets oppressively hot. He tells the crowd he is "burning up" and moves toward an onstage fan. But he does not remove his fleece jacket. Later he confesses he almost fainted shortly before going onstage.

In May 2009, Zuckerberg gained yet another powerful ally for internationalization when the Moscow-based Digital Sky Technologies spent $200 million for a small chunk of Facebook. Digital Sky, a holding company that invests exclusively in Internet companies, is the primary owner of Russian Facebook clone VKontakte ("In contact"). That, in fact, is what emboldened Managing Director Yuri Milner to make the investment. VKontakte is by far the largest social network in Russia, with a penetration of domestic Internet users beyond 50 percent, and is soundly profitable, according to Milner. Much of its sales come from virtual goods. VKontakte yields revenue more than five times what Facebook gets per user (which is less than $2 per year). "What we see," says Milner, "is that when the market is mature you can really make a lot of money on a per-user basis. If Facebook can achieve what we're seeing in Russia, that's really pretty good."

Milner's confidence that Facebook will eventually be profitable at a gigantic scale is what emboldened him to invest at a price that valued the company at $10 billion. Big as that is, it's considerably less than the $15 billion valuation that Microsoft and Li Ka-shing accepted in October 2007. Doubts lingered about Facebook's ability to be a business, and financial markets had cratered since the Microsoft deal. But Digital Sky's enthusiasm was such that not only did it buy stock from Facebook, Milner is also spending as much as $300 million more buying stock from employees and outside investors. Milner says his commitment to Facebook is long-term and that he may not sell his shares even at its initial public offering of stock, when investors frequently cash out.

Facebook's burgeoning global expansion presents challenges both technical and managerial for Zuckerberg. For one thing, Facebook's only two data centers remain in the United States and everything users around the world see on Facebook emanates from there. It can take a long time for Facebook pages to load on distant screens. That makes it even more amazing that Facebook has developed such a gigantic overseas user base. The company will have to build several very expensive additional server farms. Though it has begun opening offices, a substantial business infrastructure has to follow as well. The company has established an international headquarters in Dublin and sales offices around the world, with more to come.

Then there's the complexity of ensuring that those hundreds of millions of users and tens of thousands of application developers around the world adhere to Facebook's rules, no matter their language. For example, Facebook didn't notice that groups were talking freely in Arabic about "pig Jews" until Israeli activists pointed it out. The groups were shut down for violating Facebook's prohibition against hate speech. But it's an open question how Facebook will monitor, for example, hate groups in languages like Tamil (Tamil guerrillas waged civil war in Sri Lanka for over thirty years). So far the company is content to let users do the monitoring themselves, much as they did translation.

A provocative signal about Facebook's future arose in Indonesia in mid-2009. With 8.5 million users at that time, it had become the country's most popular Internet site. Facebook's popularity led seven hundred of the Muslim nation's imams to rule on its acceptability at a two-day meeting. "The clerics think it is necessary to set an edict on virtual networking, because this online relationship could lead to lust, which is forbidden in Islam," said a spokesman for the clerics as the meeting got under way. In their nonbinding ruling the imams said, "Facebook is forbidden" if it is used for gossiping, flirting, spreading lies, asking intimate questions, or vulgar behavior. However, overall the clerics came out surprisingly upbeat. Not only could Facebook "erase time and space constraints," they noted approvingly, but it could make it easier for couples to learn whether they are compatible before they get married. By February 2010 more than 17 million Indonesians used Facebook.

15

Changing Our Institutions "Are you familiar with the concept of a gift economy?"

One night over dinner I asked Mark Zuckerberg about Facebook's effects on society—especially politics, government, media, and business. He responded by talking about the potlatch. That's a traditional celebration and feast of native peoples on the northwest coast of North America. Each celebrant contributes what food and goods they can, and anyone takes what they want. The highest status goes to those who give the most away.

"Are you familiar with the concept of a gift economy?" Zuckerberg asks. "It's an interesting alternative to the market economy in a lot of less developed cultures. I'll contribute something and give it to someone, and then out of obligation or generosity that person will give something back to me. The whole culture works on this framework of mutual giving. The thing that binds those communities together and makes the potlatch work is the fact that the community is small enough that people can see each other's contributions. But once one of those societies gets past a certain point in size the system breaks down. People can no longer see everything that's going on, and you get freeloaders."

Zuckerberg says Facebook and other forces on the Internet now create sufficient transparency for gift economies to operate at a large scale. "When there's more openness, with everyone being able to express their opinion very quickly, more of the economy starts to operate like a gift economy. It puts the onus on companies and organizations to be more good, and more trustworthy." All this transparency and sharing and giving has implications, in his opinion, that go deep into society. "It's really changing the way that governments work," he says. "A more transparent world creates a better-governed world and a fairer world." This is, for him, a core belief.

While many would surely question Zuckerberg's idealistic notion that a more transparent world will necessarily be better governed and fairer, it's worth examining some of the effects the service is having. Zuckerberg essentially argues that any individual's public expression on Facebook is a sort of "gift" to others. That has different manifestations depending on what kind of expression it is. In the most humdrum of exchanges, when one high school student writes on another's wall, "LOL that was a funny comment," it is merely the gift of being ourselves in front of others, of including our friends in our lives. That's hardly anything new. It's just happening in a new electronic neighborhood.

When it comes to political activism, Facebook offers a more fundamentally altered landscape. In most cases we are irrevocably identified by our names there. When we say something on a political subject we are exposing our views. Others will not necessarily share them. The "gift," so to speak, is what we do for others when we put our ideas out there and make ourselves vulnerable to criticism, which can easily on Facebook be directed at us under our real names. In Zuckerberg's view, you are in essence making a gift into this free-sharing economy of ideas if you comment on Facebook about, for example, President Obama's health-care reform efforts. Think of it as a gift of opinion into the polity, a gift of ideas that may ultimately strengthen the polity.

Joining a protest group on Facebook is unlike standing in a crowd and holding up a sign at a protest. It may be easier to do in terms of convenience, but it is a more public commitment. It's more like signing a petition with our name and address in a way that many others can immediately see. Think of how Oscar Morales hesitated that last night before he took the leap of creating his group against FARC. Facebook for the first time gave him a platform where he felt comfortable taking the leap, whereas in the past in Colombia such expressions had often been considered too risky.

Our act of expression is less fraught when we are passing on an opinion about commercial behavior—telling what we think about a company or product—or when we are merely forwarding something like a news story we've seen and found interesting. Nonetheless, we are making a gesture of friendship and generosity, albeit in a way that Facebook makes routine. And that gesture potentially alters the landscape of business and media by enhancing the relative power of the consumer vis-à-vis the company or large institution. In all these sorts of beneficial expressions, you are rewarded for your contribution, typically by the reciprocal contributions of friends, and often by a sort of chain reaction of contributions by others you don't even know. Facebook is of course not the only service that enables these effects either in business or politics. Twitter, notably, is another. But Facebook is by far the largest tool of its kind.

Will Anderson, a student at the University of Florida, experienced Facebook's power after he became alarmed when he heard in early 2008 about a bill that had been introduced in the state legislature. It would redirect state scholarship money that was going to liberal arts students like him and divert it to those studying math and science. Like Morales, he took a leap. Anderson started a Facebook group called "Protect Your Bright Futures" and invited 200 Facebook friends to join. Within eleven days the group had swollen to 20,000 members. That's when Anderson received a phone call from Jeremy Ring, the state senator who had sponsored the bill. He was withdrawing it. "You can't ignore 20,000 people," Ring told the *South Florida Sun-Sentinel*.

In Egypt, demonstrators in 2009 organized on Facebook to protest a proposed law that would limit bandwidth consumed by Internet users. Shortly afterward, the minister of communications significantly amended the plan to address their concerns. In a country like Egypt, where public protest can lead to torture and arrest, such successes are especially striking. In Indonesia, a woman was arrested for the absurd "crime" of criticizing a hospital in a private email to friends. After tens of thousands joined a Facebook group complaining about this injustice, she was released from prison and the focus of attention shifted to possible malfeasance by prosecutors. These are both countries where in the past, protesting publicly under your real name was risky.

Facebook has now become one of the first places dissatisfied people worldwide take their gripes, activism, and protests. These campaigns on Facebook work well because its viral communications tools enable large numbers to become aware of an issue and join together quickly. When police conducted drug raids in late 2008 on three nightclubs in Stellenbosch, South Africa, a group on Facebook formed to protest the tactics and gained 3,000 members in thirty-six hours. Comedian David Letterman made a sexual joke about Sarah Palin's daughter, and 1,800 joined a Facebook protest page within days. (Letterman later apologized.) Citizens joined on Facebook to protest a jail expansion near San Diego; a new parking lot in Dunedin, New Zealand; a campground for gypsies in Bournemouth, England; a plan by the Philippine House of Representatives to amend the country's constitution; and the relocation to Bermuda of prisoners from the U.S. military prison in Guantanamo Bay.

"I call this digital democracy," says author Jared Cohen. A former student of Bush administration secretary of state Condoleezza Rice, Cohen was hired by Rice to join the State Department's critical Policy Planning staff. "Facebook is one of the most organic tools for democracy promotion the world has ever seen," adds Cohen. When he arrived at the State Department in late 2006 at age twenty-four, he was reluctant even to mention Facebook in meetings. People there had barely heard of it. But Facebook kept growing globally. By late 2008 it was being discussed in the White House Situation Room, where President Bush and his National Security Council staff gathered during crises.

During the waning days of the Bush administration, Cohen, Rice, and other top State Department officials took notice of what had happened in Colombia. Could Facebook, they wondered, enable people to come together and take political action even in the most repressive societies? Could it be an effective tool against terrorism? After all, Morales's Un Millon de Voces Contra Las FARC was an antiterrorist movement.

The State Department started to pay close attention to groups like Young Civilians in Turkey. This irreverent organization, whose cause is tolerance and democracy in a very diverse Muslim country, is made up mostly of students and young adults. Its symbol is a red high-top sneaker, to humorously underscore its distance from the booted military that so dominates Turkish daily life. Facebook has deeply penetrated Turkey's population—most educated young people are users. Young Civilians has 13,000 members on Facebook, which has become a primary communications tool. In a country often torn by ethnic and religious enmity, the group prides itself on including Turks of all ethnic groups and beliefs, including Kurds, Armenians, and other longtime victims of discrimination. Young Civilians uses Facebook to help organize marches where gays march next to covered Muslim women.

In December 2008, Facebook, AT&T, MTV, Google, and Net video company Howcast brought representatives of seventeen Facebook-fueled youth activist groups from around the world, including Young Civilians, to Columbia University for a two-day conference called the Alliance of Youth Movements Summit. The idea was to help protolerance and antiterrorism groups cross-pollinate and return to their countries strengthened by the exchange. Colombia's Oscar Morales came to New York and addressed the groups, as did Bush administration undersecretary of state for public diplomacy James Glassman.

"This is public diplomacy 2.0," Glassman said in a speech. "The new technologies give the U.S. a significant competitive advantage over terrorists. Some time ago I said that Al Qaeda was 'eating our lunch on the Internet.' That is no longer the case. Al Qaeda is stuck in Web 1.0. The Internet is now about interactivity and conversation. Now the Net itself is becoming the locus of Civil Society 2.0. Meanwhile, Al Qaeda keeps its death cult ideology sealed off from discussion and criticism." Then he looked out at the group of young Facebookers from Burma, Colombia, Cuba, Egypt, Lebanon, Mexico, Saudi Arabia, South Africa, Turkey, the United States, and the United Kingdom. "You are the best hope for us all," he said. He was applauding what seemed to be a new willingness to take the risk of taking a political stand on Facebook. He talked about it as a change in the balance of global power. Political activism on Facebook illustrates what foreign affairs expert Fareed Zakaria in his book *The Post-American World* calls "the rise of the rest." Nontraditional forces are gaining influence worldwide, Zakaria explains, including nonstate sources of power like those manifested in Facebook groups.

Until Facebook came along, there was hardly anywhere on the public Internet where you had to operate with your real name. In most cases anonymity remains rampant. That has often had unfortunate consequences. As Glassman said, Al Qaeda and the malefactors of the world want to remain cloaked and to avoid open discussion with their adversaries. And though it's less pernicious, think of the impulsive and often vicious anonymous comments on many blogs, or the irresponsible interactions that so often characterized behavior in AOL chat rooms. On Facebook you must have the courage of your convictions.

If you troll through groups already functioning on Facebook, it isn't hard to find examples of those that are in various ways facilitating cross-cultural understanding. Facebook has already been used, for example, to connect a global group called the Muslim Leaders of Tomorrow — 300 young Muslims from seventy-five countries, including a Saudi fashion designer, an Iranian rapper, a Pakistani madrassa reformer, an American blogger, and a Dutch lawyer. They gathered for a global conference devoted to peace and justice in Doha, Qatar, in 2009, and continue to work together as a group on Facebook.

Nonetheless, there are plenty of less friendly groups on Facebook, including those showing sympathy for Al Qaeda. So long as they do not contain explicitly hateful language or advocate illegal acts, they conform to Facebook's terms of service. Positive messages are not assured of dominating on Facebook.

While a willingness to be public about your views may be admirable, some say that it is in fact too easy to join political groups on Facebook. When you can express a view so readily, with one mere click of your mouse, the conviction behind the expression may be proportionately weaker and it's often unclear whether the number of people who join a group or cause means very much. Attempting to answer the question, three University of California at Santa Barbara political scientists published in 2009 a paper they called "Facebook Is . . . Fostering Political Engagement: A Study of Online Social Networking Groups and Offline Participation." By correlating student membership in Facebook political groups with how involved they became in the real world, they concluded that "membership in online political groups via the Facebook platform encourages offline political participation."

Politicians too can benefit from Facebook's gift economy. Barack Obama's 2008 presidential campaign used Facebook masterfully. Facebook co-founder Chris Hughes, who joined the company full time after graduating, later left to take a senior role in the campaign's online strategy team. Obama of course had a large Facebook page, which gathered millions of fans during the campaign. But in addition, local and regional Obama campaigns invited supporters to join their own Facebook groups, which allowed them to mobilize local supporters en masse.

Obama so mastered digital tools that some dubbed 2008 "the Facebook election." Nick Clemons was director of Hillary Clinton's successful primary campaign in New Hampshire and several other states. Because of Facebook, he felt at a disadvantage. "On the Clinton campaign we could definitely feel the difference because Obama was using those tools," he says. "Someone says, 'I'm going to canvass for Barack Obama,' and gets it out to thirty friends on Facebook. And if five people send it out, it multiplies. They recognized this technology earlier than anyone else, and it had a lot to do with them getting the energy and commitment of that generation of people who had not been involved in campaigns previously."

Obama remains the most popular American politician on Facebook, with about seven million supporters of his public profile as of early 2010. ("Favorite music: Miles Davis, John Coltrane, Bob Dylan, Stevie Wonder, Johann Sebastian Bach (cello suites), and The Fugees.") But number two is former Republican vice presidential candidate Sarah Palin, with more than 1.4 million.

Palin's success demonstrates that Facebook is not the preserve of any one political orientation. She has mastered the art of Facebook politics. After she resigned from her post as governor of Alaska, she began managing her public presence almost exclusively on Facebook.

In August 2009 she catalyzed national conservative resistance to President Obama's proposed health-care reforms by asserting in a post on her Facebook page that Obama aimed to create "death panels" to determine who could live or die. When the note stirred up a national controversy Palin did not respond at all until, five days later, she posted yet another Facebook post titled "Concerning the 'Death Panels.'" It got her massive coverage in the traditional media and attracted several hundred thousand new supporters. "Facebook is perfectly suited for someone as polarizing as Sarah Palin," Ari Fleischer, former press secretary for President George W. Bush, told the website Politico. "It's the ideal way for her to keep in touch, to rev up her base and go around the mainstream media." Another Facebook and Twitter master is Scott Brown, the Republican candidate who came from nowhere to win the special election in January 2010 for Ted Kennedy's Massachusetts senate seat.

Facebook has been embraced by many governments as a tool to communicate more efficiently with citizens and employees, in situations both large and small. After Hurricane Gustav hit Louisiana in early September 2008, Facebook targeted users in the affected region and used a special announcement on the top of its home page to ask them all to update their Facebook status with an indication about their safety. It coordinated this information with state and federal agencies to provide real-time data about human needs in the affected regions. It intends to use similar procedures in future disasters. In a less dire example, after thousands were denied access to Obama's January 2009 inauguration and became stranded in a Washington underground tunnel for hours, some formed a Facebook group called Survivors of the Purple Tunnel of Doom. It quickly gained more than 5,000 members. Shortly thereafter, Terrance William Gainer, the sergeant-at-arms of the U.S. Senate, who was responsible for much of the inauguration security, came onto the group's Facebook page, wrote a lengthy apology, and engaged in dialogue with some who had been trapped.

Facebook communication is becoming routine for agencies at all levels of government. When the New York City Department of Health wanted to promote the use of condoms to prevent the spread of HIV, it launched a Facebook page and application that allowed users to send one another a little image of a so-called "e-condom." The commandant of the U.S. Coast Guard updates his Facebook status using his cell phone when he travels, and the top U.S. general in Iraq maintains a Facebook page to answer questions about U.S. activities there. The White House streams President Obama's press conferences on Facebook, enabling users to comment in real time with one another next to the event. Even the Saudi Arabian minister of information has created a profile on Facebook, where he accepts journalists as friends, takes their interview requests, and releases information. Now government leaders in many places are starting to talk about making it possible to renew driver's licenses and interact in other ways with government on Facebook.

Facebook is the biggest of a number of websites redefining news into something produced by ordinary individuals and consumed by their friends. I create some news for you, you create some news for me— Zuckerberg's gift economy again.

When Thefacebook first launched at Harvard in 2004, on each person's profile page was a list of all the articles from the archives of the *Harvard Crimson* in which he or she was mentioned. The feature was quickly removed. In a 2009 post for the Nieman Journalism Lab, Zachary Seward, a student at Harvard back then, suggests "Zuckerberg . . . realized that Facebook wasn't a tool for keeping track of news made somewhere else. It was a tool for making news right there, on Facebook." And that is in fact exactly how Zuckerberg has always viewed the News Feed—a real source of relevant news, both about your friends and about the world. Long before Facebook debuted News Feed in 2006, Zuckerberg had meticulously articulated in his diaries exactly how its updates would be real news, going so far as to create a style sheet and grammatical rules for News Feed "stories."

News on the News Feed was far more personal than what any professional media organization had ever attempted to deliver. It was ordinary everyday information about what your friends were doing and what they were interested in. Recall the rationale Zuckerberg gave internally for the News Feed: "A squirrel dying in front of your house may be more relevant to your interests right now than people dying in Africa." Now your every move on Facebook might become news for your friends.

On campuses, the near-total penetration of Facebook at U.S. high schools and colleges has rendered traditional campus print media—the newspaper and the yearbook—far less urgent. People find out what's going on and who's doing what on Facebook. It's possible that focusing on this diurnal news may make people care less about serious events more distant from them—those people dying in Africa, for instance. It's one of many important Facebook-related social questions that deserve further study.

Sean Parker, who helped Zuckerberg develop his basic views about the service, is passionate about Facebook's importance in altering the landscape of media. In his view, individuals now determine what their friends see as much as the editor at the local newspaper did in simpler times. Facebook permits your friends to, in effect, construct for you a personalized news portal that functions somewhat like the portals of Yahoo or AOL or Microsoft. If I see a friend post a link to something in a field I know they're expert in or passionate about, I am more likely to click it than I am to click something that shows up on my MyYahoo home page. And in the inadvertent spirit of a gift economy, in return I frequently post links to things I find interesting, useful, or amusing. The ever-intellectual self-educated Parker calls it "networks of people acting as a decentralized relevancy filter." A similar but more anonymous form of sharing is facilitated by websites like Digg, Reddit, or Twitter.

If a message is powerful enough it can spread to a vast sea of connected individuals, regardless of who originated it. Chris Cox, Facebook's vice president for product and a close Zuckerberg protégé, puts it this way: "We want to give to everyone that same power that mass media has had to beam out a message." The leveling of the playing field is much in evidence. For example, it was via Facebook status updates that newscasters at CNN first learned of the January 2010 earthquake in Haiti, a network executive said that day on the air.

So how do traditional media organizations fit into this new person-centric information architecture? Paradoxically, if they are to most benefit from the Facebook environment they have to learn to function within it as if they were individuals. The playing field has been leveled by the site's neutral way of treating all messages as similar. Any media company, newspaper, or TV station can create its own page on Facebook. But then it faces the same mandate to generate interesting, relevant, and useful messages that an individual does. Activity on a page gets deposited into users' News Feeds—just like the activity on any individual's profile. First you have to get someone to embrace you as your "fan," much like becoming a "friend" of an individual. Then the goal is to get people who see the information you produce to endorse it themselves by clicking Facebook's ubiquitous "like" indicator or by commenting on what you post. That forwards it further to their network of friends and keeps it virally alive. Largely because of the efficiency of this process, Facebook has become one of the top drivers of traffic to major media websites, often behind only Google. Facebook may also challenge conventional media financially over time—by, along with other websites, drawing away the lucrative brand advertising that has been a mainstay of TV, magazines, and newspapers.

Facing these changes, many major media companies are trying to work with Facebook rather than against it. NBC, for example, in summer 2009 previewed an upcoming new series called *Community* exclusively on Facebook. Only those who identified themselves as the show's fans could see the preview episodes. NBC advertised on its own website as well as on Facebook that these previews were available. The service's penetration among the young and media-savvy demographic presumed to be the show's audience meant that the preponderance of potential viewers were already on Facebook. So limiting it to Facebook didn't limit the audience so much as it provided information about exactly who the audience was, since Facebook can provide aggregate demographics of a page's fan base to companies.

The line between Facebook and old media is blurring. Verizon has incorporated Facebook along with Twitter and a few other social media websites into its FIOS broadband television rollout. You can log in to Facebook on your TV using your remote, and on a split screen use it to update your status and share information with friends about shows you're watching. Some media companies, like the Huffington Post, have deeply integrated Facebook into their websites so users can use their Facebook identity to share and comment on stories and videos with friends.

The next phase is likely to be a more thorough marriage between Facebook and conventional media, especially television. As the FIOS integration suggests, Facebook gives viewers a platform to in effect watch TV with their friends. There are other ways to do the same thing. Facebook has also made it quite easy for any video broadcast on the Web to be accompanied by live commentary by Facebook users through their status messages, which can be seen on any site's page that chooses to integrate them. One of the first examples of such integration was when CNN enabled users to comment online during the inauguration of President Obama. You could watch the updates of all the other viewers (which reached 8,500 per minute) or just those posted by people on your own friend list. ABC.com did something similar during the 2009 Academy Awards.

A world in which each individual has a clear window into the contributions of everyone else, potlatch-style, does not dovetail well with how most companies are run. While employees of just about every company in America are on Facebook in force, its intersection with the classically structured corporation has been awkward and clumsy so far. Gary Hamel, one of the great theorists of modern management, considers that inevitable. "The social transformation now happening on the Web," he explains, "will totally transform how we think about organizations large and small." Hamel says historically there have been only two basic ways to, as he puts it, "aggregate and amplify human capabilities." They were bureaucracy and markets. "Then in the last ten years we have added a third—networks. That helps us work together on complex tasks, but it also destroys the power of the elite to determine who gets heard."

Few companies have wrestled effectively with this contradiction. Elites—such as the managers of the typical corporation—seldom willingly surrender power and authority. Says author and strategy consultant John Hagel: "Companies are facing the same issues that individuals are facing, which is the degree of transparency and openness that's appropriate," he says. "But in general individuals are moving more rapidly and developing more appropriate social practices than institutions are." This is one of several reasons why many companies now restrict the use of Facebook in the office. The spread of Facebook as a communications medium so far has been too rapid for most managements to have understood what it means.

Some executives, however, have embraced Facebook in the enterprise. When they do they almost universally encounter social dynamics that unsettle the corporate power equilibrium. At Serena Software, a Silicon Valley company that was running out of gas as a provider of software for mainframe computers, new CEO Jeremy Burton turned to Facebook in late 2007 as a tool to shake up a hidebound, old-school corporate culture. Serena even set aside a couple of hours weekly on what it called "Facebook Fridays" for employees to establish Facebook connections with co-workers, suppliers, customers, and anyone else.

Burton became Facebook friends with hundreds of Serena's nine hundred employees. As a result, Burton gained useful insights into how Serena functioned day to day. Employees casually posted details about their jobs and sent him surprisingly candid Facebook messages. "People feel more comfortable telling the CEO things on Facebook than they ever would in person or with email," he says. "They feel it's more informal." But informality comes with other costs. Burton's much younger brother in England sometimes bluntly disagreed with what Burton said on Facebook, in full view of employees and other friends.

Then came 2008's precipitous economic downturn. Serena, like every other company, saw revenue plummet. Burton had to lay off about 10 percent of the company's workers. Accordingly he had to decide whether once an employee was laid off he ought to "unfriend" the person on Facebook. He found the layoff process deeply unsettling, and shared some of his feelings about it on Facebook. A couple of people who were laid off sent him sympathetic notes there, acknowledging the challenges he faced or proclaiming their time at Serena to have been valuable no matter how unhappily it ended. He remained Facebook friends with several people he fired.

At a completely different kind of company, global journalism and financial information powerhouse Thomson Reuters, Editor in Chief David Schlesinger found a similarly informal dynamic. He's a rabid Facebook partisan who checks the service "easily two dozen times a day," and who, as manager of one of the world's biggest news services, concedes, "I actually think the Facebook News Feed is real news. It tells me news I'm interested in." He mostly uses Facebook to connect with colleagues and employees but says the way he relates to people there does not depend on where they work. "There are some journalists six levels below me in the hierarchy with whom I have a very intimate relationship on Facebook," he says. "A junior reporter who is my friend may ask me for advice on a story when they would never dare do it by email or telephone or in person. It's wonderful. I love it. The HR jargon for it would be level-jumping." Schlesinger, like Burton, is a secure manager who wants to empower people in his organization. Executives more eager to exercise power themselves will not find it so comfortable. Most of them—and we all know how many there are—stay off Facebook.

Companies are often eager to get marketers and sales executives onto Facebook as its importance in that world grows. Sony Pictures, an early Facebook advertiser, decreed back in 2006 that executives should have Facebook profiles. At computer-chip maker Intel, the sales and marketing departments conducted a sort of treasure hunt with an iPod as the prize. To participate you had to start with clues at a fictitious Facebook profile. But in order to see that profile you had to create one for yourself.

From early on, companies have been approaching Facebook asking for special features for enterprise use, but Zuckerberg has never been particularly interested. Companies want, for example, to be able to sequester employee conversations so absolutely no outside "friends" could ever see their internal discussions. That remains impossible. Executives at Facebook say such capability will eventually get built, it's just not a high priority now as the company is growing so quickly among consumers. But co-founder Moskovitz feels strongly about building features that help companies collaborate internally in the way that Facebook has made it so easy to "collaborate" with your friends. The presumption at Asana, Moskovitz's San Francisco–based start-up, is that electronically facilitated collaboration will increasingly be built into the fabric of every successful enterprise. At Facebook, Moskovitz consistently advocated giving employees tools that empowered them inside the enterprise, and many of his innovations remain in use there today.

Microsoft, the world's leading business software company and a big Facebook investor and partner, has periodically campaigned to get Facebook to enable a version of its service to work in conjunction with Microsoft Office. That idea has consistently been met with yawns, to the consternation of some at Microsoft. Now Salesforce.com, a smaller but agile competitor of Microsoft, has launched a social network for businesses called Chatter. Companies of many types are beginning to experiment with that and similar products.

Facebook itself is both a beneficiary and a victim of the dynamics of the gift economy its CEO is so partial to. The more users want to contribute, the more activity they generate and the more page views Facebook can use to display advertising. But because Zuckerberg has given Facebook's users such powerful tools to express their views, the company itself has regularly borne the brunt of user dissatisfaction when it took actions people disapproved of. Digital democracy affects life inside Facebook even more than outside it.

Zuckerberg accepts this as inevitable. "We're a vehicle that gives people the power to share information, so we are driving that trend. We also have to live by it," he says. That was tough enough to deal with in the quaint days of the News Feed controversy, when Facebook had fewer than 10 million users. Now, with the burden of more than 400 million empowered and contributing users, Zuckerberg's life is becoming considerably more complicated thanks to the extraordinary tools he has made available to all these people.

16

The Evolution of Facebook "What we're doing now is just the beginning."

On the first workday of 2009, Mark Zuckerberg—he of the rubber sandals, T-shirts, and fleece jackets—arrived at work wearing a conservative tie and a collared white dress shirt. "It's a serious year," he told everyone who asked. He was going to wear a tie all year, he explained, to underscore the issues Facebook faced as growth reached stratospheric levels.

But it wasn't growth per se that made Zuckerberg feel he needed to signal a new seriousness to his peers. It wasn't the need for "monetization," either. Rather, it was the challenges that come with being a rapidly evolving communications platform that has already been embraced by a mass audience.

Zuckerberg still sees Facebook as a work in progress. Toward the end of 2008 I asked him what he considered its biggest challenge. "The biggest thing is going to be leading the user base through the changes that need to continue to happen," he answered without any hesitation. "Whenever we roll out any major product there's some sort of backlash. We need to be sure we can still aggressively build products that are on the edge and manage this big user base. I'd like us to keep pushing the envelope."

Facebook was still less than five years old, but it had already brought its users through a series of major changes. The inclusion of photos, the introduction of the News Feed, and the expansion of Facebook through the applications platform and the translation tools had each, in its own way, fundamentally altered the product and transformed the user experience. Now Zuckerberg and his engineers were planning further dramatic changes. He wouldn't think of abandoning them. It was going to be a serious year.

. . .

Even in late 2008, when Zuckerberg confessed these concerns about keeping Facebook moving forward, he had already initiated a series of changes intended to get users exchanging even more information with one another. In September 2008, only two weeks after briefly celebrating 100 million active users, with a toga party, Facebook reorganized profile pages in a way many found jarring. As always it led to loud user protests. Inside the company the initiative was nicknamed "FB 95" in an ironic and admiring wink to Windows 95—the Microsoft operating system that finally and indisputably made Windows a mass-market product and turned Windows-based PCs into a resilient worldwide monopoly. This change to profile pages was supposed to similarly help Facebook blanket the world.

The primary aim of the redesign was to increase the velocity of information flowing between users—or "sharing," in the lexicon of Facebook—and to simplify the site's design to make it easier to digest an ever-increasing volume of information. In the most significant change, two separate components of your profile were combined—the "wall," where friends sent you direct public messages, and the "mini-feed," the personalized News Feed that displayed information about you. Now everything that was about you was in one place. A central aim was to create more launchpads for discussion. At the top of your profile was now a box called the "publisher"—an enhanced version of the old slot where you merely posted status updates. But the box was now for content of all types, everything from quotidian updates of the classic sort—"I'm getting into the shower now"—to photos, videos, and links to articles and sites of interest around the Web. Whereas Facebook's old status update box had prompted you with something like this "David Kirkpatrick is . . . ," now the publisher box included a much more open-ended question: "What's on your mind?"

In order to ease Facebook's increasingly skittish users into the new design, the company gave users a trial version almost two months before requiring them all to shift over. It maintained old and new versions in parallel. As Zuckerberg said, "The technology is the least difficult part." Managing Facebook was becoming an exercise in crowd psychology.

But careful user relations only went so far. Many users hated the redesign. Thousands again joined groups protesting it, though not nearly as many as had protested against News Feed. A few days after the redesign, even computer executive Michael Dell joined a group called Petition Against the "New Facebook." Young people especially were attached to their old wall, which had been in place in one version or another since late 2004.

On the day in July 2008 when Facebook first showcased its redesign, influential tech journalist Michael Arrington wrote a prophetic item on his widely read TechCrunch news site. It was titled "The Friendfeedization of Facebook." FriendFeed was a small website started in October 2007 by several former top Google engineers. As Arrington pointed out, it "expertly combined the idea of an activity stream that was first popularized by Facebook with the microblogging trend introduced by Twitter." Now, with its redesign, Arrington saw Facebook mimicking FriendFeed by taking its own traditional News Feed content and blending it with beefed-up status updates that resembled the so-called tweets on Twitter.

For the first time since it emerged, Facebook was now being forced to react, at least in part, to the innovations of others. And while it may have begun to look a bit like still-tiny FriendFeed, the major new force in the equation was Twitter. Created in 2006, Twitter gives users a forum to post updates of no more than 140 characters. To many, especially people who don't use both, Twitter seems much like Facebook, because both put great emphasis on rapid sharing of information between individuals. But on Twitter people do not become "friends." Instead you can sign up to "follow" anyone's tweets—the name users give its telegraphic updates. Twitterers are not necessarily even people. A large percentage of Twitter accounts use aliases or company names. And unlike those on Facebook, Twitter connections are one-way. Facebook's heritage is as an identity-based platform to communicate with people you know offline, but Twitter is a broadcast platform—a medium perfect for companies, brands, bloggers, celebrities, and anyone who has something they want lots of people to know about.

There are undeniable parallels between the two products. The status update is a central feature of both. Twitter, like Facebook, opened itself up early as a platform for other applications. Indeed, many users tweet and view the tweets of others on independent sites like Tweet-Deck. Twitter one-upped Facebook as well in its blasé approach to revenue—in 2009, three years after it was founded, it still had virtually none. Growth was its mantra, and it was getting plenty of it.

Twitter's momentum with users continued to build over the subsequent months. Facebook was now large, established, and from the press's point of view, a bit old hat. Twitter was the next thing. It quickly became the "it" tech company, a status Facebook had occupied for most of 2007 and 2008. Predictions that Twitter would supplant Facebook were rife. Zuckerberg and his team were following Twitter closely. They were extremely focused on the degree to which the enthusiasm of the press and Silicon Valley cognoscenti had migrated to Twitter.

At an onstage interview at the Web 2.0 conference in early November, two months after instituting Facebook's redesign, Zuckerberg said he was "really impressed" by Twitter and called its service "an elegant model." Around that same time Facebook got deep into secret talks to buy Twitter—reportedly for $500 million in stock. The deal didn't happen, among other reasons because Twitter's executives were not confident in the potential value of Facebook's stock.

Facebook made yet another huge transition in late 2008. Zuckerberg aimed to start embedding Facebook into the very fabric of the Internet. In a fundamental change to its platform, the company launched Facebook Connect. The launch was an appeal to developers to start building on top of Facebook in a new way.

Connect makes it possible for any site on the Web to allow you to log in using your Facebook account. That accomplishes several things. It lets you bring your identity with you wherever you go online. Because you can tell Connect to send information back into your Facebook feed, it's a way to project information about the actions you take on those sites back to your Facebook friends just as if they were actions inside Facebook. It also enables Facebook to lend its virality—the way it so efficiently transmits information from one user to many friends—to any website that wants to take advantage of it.

For users, Facebook Connect offers what could turn into a universal Internet log-in. Over 80,000 websites use it in some fashion, as of February 2010, and 60 million Facebook members are actively employing it. Connect partners include about half of all the top 100 websites in the world, as measured by the comSource research firm, Facebook executive Ethan Beard told a conference audience. They range from Yahoo, the world's largest content website, to big media sites like CNN, the Huffington Post, Gawker, and TechCrunch, hot start-ups like Fanbase and Foursquare, and devices like the iPhone and the Xbox gaming console. "We aspire to be a technology that people use to connect to things they care about no matter where they are," Beard told the conference. (Remember how proud Zuckerberg was, way back in the fall of 2003, when he said that with CourseMatch "you could link to people through things"?)

When readers log in to comment or interact on one of these sites or devices using Facebook Connect they are identified by their Facebook photo and real name. This addresses a huge problem that has afflicted blogs and news sites—the significant percentage of posts by readers that have been extreme, insulting, and anonymous. When discussants log in under their real names with Connect, the dialogue becomes more civilized.

"Facebook Connect is the future of the way that platform is going to work," says Zuckerberg. "I don't think it's going to be these little applications inside Facebook. It will be whole websites that just use people's information from Facebook in order to share more information." Now he says that Facebook's internal platform, which enabled applications to operate inside the bounds of the service, was merely "an important training step."

Despite widespread enthusiasm for the opportunities Connect offers to tap into Facebook's hundreds of millions of users, some potential partners are skeptical. "It's a Trojan horse strategy," says the CEO of one New York–based media company who pays close attention to Facebook but has no intention of deploying Connect. He sees it as a method to get between him and his customers. He predicts that once Facebook makes sites dependent on its log-in and access to users, it will start making demands. For now there is no charge to use Connect, but he expects that to change.

Connect will also most likely become a vehicle for delivering advertising. This possibility has been downplayed by executives thus far. But Dustin Moskovitz, who speaks more freely now that he's left the company, says sites that use Connect will ultimately be able to display ads provided to the site by Facebook. "[They] will know which Facebook user is on their site," he explains, "so [they] can use all of Facebook's ad-targeting information. That's absolutely core to the Connect strategy." Sharing in the revenue that these targeted ads make possible on other sites could become an important business for Facebook.

Another function of Connect is that it will give Facebook even more information about users, including data no longer limited just to what they do on Facebook.com.

In January, around the time Zuckerberg was donning his tie, a potentially serious internal crisis erupted at Facebook. As President-elect Obama was assembling his cabinet and advisers, he hired Lawrence Summers to be the chairman of the National Economic Council at the White House. When Summers had been secretary of the Treasury under Bill Clinton his chief of staff had been Sheryl Sandberg. Summers and Sandberg have remained close, and some senior people at the company worried she might join the new administration and thought it a real possibility. She decided to stay put. She was becoming an essential partner to Zuckerberg.

In February, the year got even more serious. Facebook's legal department posted a few changes to the company's "terms of service," the legalese that is intended mainly to indemnify a company against lawsuits by disgruntled users. This new version of the rules, which every new user must stipulate they have read and agreed to, even though they usually don't, were at first ignored by virtually everyone. But at 6 P.M.

on Sunday, February 15, a blog called the Consumerist, published by Consumers Union, took a close look at the changes and published a post titled "Facebook's New Terms of Service: 'We Can Do Anything We Want With Your Content. Forever.'"

The article expressed alarm about the terms and quoted a section about what happens to content you post: "You hereby grant Facebook an irrevocable, perpetual, non-exclusive, transferable, fully paid, world-wide license (with the right to sublicense) to . . . use, copy, publish, stream, store, retain, publicly perform or display. . . ." Actually, that terrifying-sounding language was unchanged from the previous version, but in a key change, a subsequent clause had been excised. It said that if you removed your content from Facebook, this license would expire. Removing that clause changed everything, in the opinion of the Consumerist. Its recommendation: "Make sure you never upload anything you don't feel comfortable giving away forever, because it's Facebook's now."

This post was quickly picked up by a number of other blogs and by many in the mainstream press. Suddenly Zuckerberg was under unexpected pressure. How, asked a swelling volume of articles appearing around the world, could he assert he owned the information Facebook's users posted there? He couldn't. And in his opinion, he hadn't. But unlike in some earlier incidents, he was prepared to say so immediately. By 5 P.M. Monday he had posted a lengthy and thoughtful response on the Facebook Blog, titled "On Facebook, People Own and Control Their Information." "In reality, we wouldn't share your information in a way you wouldn't want," Zuckerberg wrote, attempting to reassure users. But then he went on to explain the complicated new legal terrain that a service like his now operated in. Users want to control their own information, but they also want to control and sometimes move information other users have entrusted to them—such as cell-phone numbers, photos, etc.

It was not enough. A twenty-five-year-old user from Los Angeles named Julius Harper quickly created a group called People Against the New Terms of Service, which soon merged with another protest group created by Anne Kathrine Petteroe of Oslo, Norway. By Tuesday the group had 30,000 members. By Wednesday it was 100,000. Again the tools for rapid communication and organization that Facebook gives its users were being deployed against it. Meanwhile, the Electronic Privacy Information Center and twenty-five other consumer protection organizations were preparing to file a complaint with the Federal Trade Commission on Wednesday.

Zuckerberg quickly surrendered, less than three days after the original article appeared. At 1 A.M. on Wednesday, he announced on the blog that Facebook was temporarily reverting to the old terms of service while it decided what to do next. He had said even in his earlier note that he agreed that much of the language in the terms seemed overly formal and needed to be simplified. In this late-night note, he invited Facebook's users to join a newly formed company group to discuss what the terms ought to say, and promised "users will have a lot of input in crafting these terms."

The following week Zuckerberg announced that Facebook had created two new documents: a set of Facebook Principles to lay out the "guiding framework" for company policies, and a "Statement of Rights and Responsibilities" that would replace the old terms of service. He asked people to comment on both, and announced that users would be invited to vote for or against them before they went into effect. He ended with a kind of rhetoric you seldom hear from CEOs: "History tells us that systems are most fairly governed when there is an open and transparent dialogue between the people who make decisions and those who are affected by them. We believe history will one day show that this principle holds true for companies as well, and we're looking [forward] to moving in this direction with you."

In subsequent weeks Facebook lived up to its pledge. It invited the creators of the original protest group, Harper and Petteroe, to help it evaluate and organize comments about the documents. Zuckerberg announced a vote that would be binding if at least 30 percent of Facebook's users participated. Since, the week before, he had announced that Facebook now topped 200 million active users, that meant 60 million people would have to vote, an unlikely prospect. But he was at least in theory submitting to the will of the people.

In the end only 666,000 votes were cast, with 74 percent of users favoring the revised Statement of Rights and Responsibilities. The Consumerist pronounced itself satisfied. Internet activists were impressed. Jonathan Zittrain, a professor at Harvard Law School and author of the alarmist book *The Future of the Internet—and How to Stop It*, wrote an admiring article noting that Zuckerberg had encouraged Facebook's users to view themselves as citizens—of Facebook.

Zuckerberg was pleased when I talked to him two weeks after the results were announced. He planned more such votes in the future. "If we do something controversial, what this will really mean is that we're accountable to our users," he told me. "We now need to communicate with them clearly about it. I think that keeps us honest." It was a serious year, but he was displaying a seriousness to match it.

In March 2009, Facebook made yet another dramatic set of changes, this time explicitly aimed at co-opting Twitter. The changes in this redesign were most visible not on your profile, with its "wall," but on the home page where you first land in Facebook and see information about your friends. The top of that page now sported a publisher box just like the one on your profile. The message was getting louder—Share! Beneath this box, the News Feed had morphed into what Facebook now called a "stream," a continuous list of updates and other information from friends. But the stream also included updates from a new source: pages where you had become a "fan." Now becoming the fan of a commercial Facebook page was almost identical to following a person or company on Twitter.

The new streamed News Feed differed from the old one in two fundamental ways. It was updated in real time (like Twitter), and it was not based on an algorithm (neither was Twitter). The old News Feed depended on software that watched your past behavior and attempted to guess what you would be interested in. You could never be sure what would surface. The new stream, by contrast, was what the eggheads at Facebook loved to call "deterministic." You determine exactly what appears there. Facebook added filters on the left side of the home page to help you control what appeared in your stream. You could use them to see videos, or photos, or status updates, for example. You could also put friends and pages into groups to create different personalized views of the stream. Now, for instance, you could see just family members, or people from your high school class, or employees at Facebook, or your best friends.

It was a heady and confusing mix. There remained a small algorithmic section on the home page called Highlights, an unappealing list of small items and tiny photographs on the lower right side of the page. Few people found it very useful. And this time Facebook abandoned the previous redesign's deliberately gentle introduction process. There were no trial periods or parallel versions to ease users into the changes. But it was immediately apparent that many of Facebook's 175 million users did not like the changes.

Nor did the company's increasingly defensive staffers expect them to. As soon as Facebook began to roll out the new design someone created a satirical group entitled I AUTOMATICALLY HATE THE NEW FACEBOOK HOME PAGE. Many of the group's members worked at Facebook. Its description read: "I HATE CHANGE AND EVERY-THING ASSOCIATED WITH IT. I WANT EVERYTHING TO RE-MAIN STATIC THROUGHOUT MY ENTIRE LIFE." Staffers posted facetious remarks. "BRING FACEBOOK BACK TO ITS FORMER GLORY. HARVARD-ONLY," wrote one. "I will hate this redesign until another iteration, in the event [of] which I will love this redesign and vehemently oppose the successor," wrote another, sarcastically.

Two weeks after the redesign, yet another Twitter-like feature was added—new privacy settings that enabled you to open parts or all of your profile to everyone on Facebook. And in what would be the coup de grâce, plans were in the works to enable users to "fan" individuals. Adding such asymmetric connections for individuals would more or less complete Facebook's mimickry and make it possible to function essentially as if you were on Twitter. But while Zuckerberg originally planned to add this feature in June 2009, he still hadn't by February 2010.

By mid-2009, Twitter had 50 million members, and Facebook continued running scared. "Every time I hang out with a Facebook employee, they ask me what I think of Twitter," Moskovitz told me in May. One thing even he worried about was that top engineers were starting to choose to work at Twitter rather than Facebook (or his own new start-up). "At Facebook we feel like if we address this, we can definitely win," he said, "but we would certainly feel like shit if we just like weren't paying attention and Twitter did something we didn't understand and got past us." Facebook board member Marc Andreessen, who is also an investor in Twitter, told me around the same time that the two companies were "elephant bumping." "It's too late for somebody to compete with Facebook on Facebook's turf," he said. "So when the threats come, they will be disruptive in nature, right? Disruptive threats tend to come up from below. They fly up your tailpipe, instead of coming straight at you. So Twitter is the kind of thing that Facebook should be very aware of."

Sean Parker, who tries hard to stay involved in Facebook product decisions from a distance, was a longtime advocate of turning News Feed into a stream that looked more like Twitter. Zuckerberg in fact resisted it for a long time, but the growing competitive pressure from Twitter, along with relentless politicking by Parker and others like Adam D'Angelo, finally convinced him. "Mark always told me he wasn't going to do it," says Parker, "but in classic Mark style he listens and listens and listens and then at some point comes to the conclusion on his own that this is the way it has to go."

Facebook's longtime self-definition as a place to connect with people you know in the real world is becoming slowly but steadily less central. To be a "friend" requires a bidirectional interaction. Both you and your friend must accede to it, as Parker explains. But now there are other sorts of useful relationships in Facebook. Parker predicts Facebook will, over time, formally separate the three components to becoming a friend with someone on Facebook—declaring that you know them, giving them permission to see your own information, and subscribing to see all the information they produce.

Zuckerberg concedes that "the concept of a 'friend' is definitely getting overloaded." He says that word was useful to "get people over a bunch of hurdles." Most importantly it got them used to sharing a lot of information about themselves—after all, only friends would see it. But

Facebook has offered only a binary choice for your relationships with others: friend or nonfriend. It will gradually offer more subtle ways to tune interactions with people. Friending will become more nuanced to more accurately reflect the various degrees of connection we have with people. All of us who have agonized over whether to accept a friend request from somebody we barely know will have more options.

But something else is going on as well—over time Facebook will become about much more than friendship. The first indication of that was when it added fan pages and sent page updates into your News Feed alongside updates from individuals. Ethan Beard, a Google veteran who is now marketing boss for Facebook's platform and a key member of Zuckerberg's team, explains: "As we've continued to evolve our thinking we've realized there is more to the graph than just people—the objects, items, organizations, and ideas you are connected with. Anything. By mapping out all these things we can come up with an extremely robust sense of a person's identity." In other words, the fact that you're a fan of U2, a coffee shop in your neighborhood, and Ayn Rand says more about you than the fact that you friended someone you met at a conference last year.

The future of Facebook will involve giving people tools to uncover relationships with other people that are manifested through common interests and behavior. Such a new direction poses the risk that it could make Facebook feel more like a place for marketing and less like a place for friendship.

As Facebook maps out all these additional connections and monitors every user's interactions with them, Zuckerberg predicts users will be sharing an ever-increasing volume of data. "Think of it as just this massive stream of information," he says. "It's almost the stream of all human consciousness and communication, and the products we build are just different views of that. The concept of the social graph has been a very useful construct, but I think increasingly this concept of the social stream—the aggregate stream for everyone—will be as important."

When he thinks about the evolution of this stream, Zuckerberg makes a comparison to Moore's law, the prediction by Intel's Gordon Moore back in the 1960s that the number of transistors that could fit on a computer chip would grow exponentially over time. He thinks there is a similar exponential phenomenon at work in social networking. In a decade, he believes, a thousand times more information about each individual member may flow through Facebook. This hypothesis has corollaries he finds intriguing. Says he: "People are going to have to have a device with them at all times that's [automatically] sharing. You can predict that."

In urging Facebook's users to turn more of their updates and other contributions into public broadcasts, and by attempting also to shuffle in their commercial behavior as well as their interactions with friends, Zuckerberg is gambling that people will over time care progressively less and less about privacy and that they will actually want all the additional information that will be coming their way. It's not just the increased volume of information that's potentially problematic. Will people tolerate so much information about themselves getting loose on the Net? With a considerable portion of the world's population on board, Facebook may become a giant experiment in personal disclosure. Zuckerberg says he remains committed to giving people the privacy controls they want. Whether he can resolve these contradictions as he changes the software beneath more than 400 million users will be fascinating to watch.

In late April 2009, Facebook quietly made a change as radical as any it had ever attempted. With the release of something called the Facebook Open Stream API, the company laid groundwork that could transform the way people use its service. The Stream API is a sort of companion to Facebook Connect. If Connect is a way to extend Facebook's platform across the Web, the Stream API represents a way to distribute the experience of being on Facebook outside the service itself. That may sound odd. Today we mostly take for granted that the way users consume Facebook information is on their home page at Facebook.com.

The Stream API lets any site take that feed and publish it elsewhere—even potentially to alter and add to it in a way that could not happen inside Facebook. It will let other services build sites that look and feel much like Facebook itself, even though the data flows will still be controlled from Facebook's servers. If I want, I could build my own website where any Facebook user could see their entire News Feed. Users can act at these external sites much as they can inside Facebook. Data can flow back into friends' News Feeds, too. The software service called Tweet-Deck enables this already, among others.

Just two days after the Stream API announcement, I had dinner in New York with Sean Parker, who spent a good portion of our time together that night denouncing it. "It's the greatest strategic gamble the company has ever made and will ever make," he said in his intense rapid cadence. "Opening the stream to the world has the possibility of breaking the company's network effect. As a closed network the switching costs are incredibly high and everybody's forced to play in Facebook's sandbox. But when you open the stream to the world you open the possibility of better Facebook clients that can process all the same data that Facebook itself can."

These words were still ringing in my ears a week later when I sat alone with Zuckerberg for a long interview in a corner conference room near his desk in Palo Alto. He didn't dispute what Parker said, but nonetheless remained unperturbed. He launched into a discussion of the perils that come when companies "build walls around themselves." "The best thing we can do is kind of move smoothly with the world around us," he continued, "and to have constant competition, not build walls. To the extent that we think most of the sharing is going to happen outside of Facebook anyway, we really want to encourage it. I can't guarantee we'll succeed. I just think that if we don't do this then eventually we will fail."

I asked him if he didn't worry that such conceptual boldness could jeopardize the company's finances. "It's only the right argument if you're trying to build something that has value over decades," he said. "It's really important for people to understand that what we're doing now is just the beginning." Chamath Palihapitiya, whose job at Facebook is to think about growth, says, "Mark has the most long-term perspective I've ever seen. This guy is uber uber uber on the long-term view." Facebook's ambitions for Connect and the Stream API remain vast. Says top Facebook designer Aaron Sittig: "If we open things up slowly over time, we could get to total ubiquity."

The company's senior managers are consistent in saying that Facebook will evolve beyond being just a "site." Its services will be available widely. It will become a storehouse for information, like a bank, but also a clearinghouse and transit point, like the post office or the telephone company. It could become nothing more than an identity registry and a hub for conveying data between different people. But that could be a very powerful position in the business ecosystem.

Some managers say Facebook on the Internet could ultimately become like an Intel chip inside a PC—something you use but seldom think about. And Matt Cohler, departed from Facebook but still deeply involved, says, "In five years there won't be a distinction between being on and off Facebook. It will be something that goes with you wherever you are communicating with people."

Think of it like having a piece of software that in effect contains your friends—or at least a persistent and potentially live connection to any of them. That "software" will allow you to stay abreast of whatever they do, and tell them whatever you want about yourself. Anytime we are doing anything online and have a question we will be able to turn to our friends. We may be able to converse with them in real time as well, through chat, voice, or video.

This experience will increasingly go with us as we traverse the real world as well, since most people will carry devices with an always-on connection to the Internet. Facebook's iPhone, BlackBerry, and Google Android applications as well as those on other mobile phones are already used by more than 100 million users worldwide. In some countries this is already the primary way people use Facebook. In the future, the main way people will use Facebook will be on a mobile device.

Here's a possible scenario: Imagine you're at a football game and your mobile device shows you which of your friends are also in the stadium—perhaps even where they're sitting. Maybe it could tell you who in your section of the stands has attended exactly the same games as you in the past. Or who is a fan of the same teams as you. This may seem cool to many users. To others it may feel Orwellian.

Shopping is another arena that could be transformed. Wouldn't you want to know, whenever you were considering an expensive purchase like a car or a refrigerator or a camera, exactly which of your friends had purchased—or maybe just considered—the same purchase? Some developer will probably figure out how to get Facebook to tell you that. It will also have to ensure that all that information flies around only with the user's consent.

Facebook might even begin to function as a sort of auxiliary memory. As you walk down a street you could query your profile to learn when you were last there, and with whom. Or a location-aware mobile device could alert you to the proximity of people you've interacted with on Facebook, and remind you how. The software could even start to make elementary decisions on your behalf. Platform marketer Ethan Beard says you will probably be able to simply tell your TiVo to record whatever shows your friends are recording. And here's a scenario he suggests: "Imagine I can get in my car and just say, 'I want to go to David Kirkpatrick's house.' It knows who I am and can go inside Facebook, find out where David lives, and direct me there using GPS. Ideas like that are so powerful, how could they not happen?"

In early August 2009, Facebook acquired FriendFeed for about $50 million, by far its largest acquisition. It really was the FriendFeedization of Facebook. Bringing into Facebook both FriendFeed's technology as well as its ex-Google founders, star coders that they are, was intended to significantly bolster Facebook's ability to compete with Twitter.

In keeping with Facebook's more elastic conception of itself, in September it launched Facebook Lite. It was the first true brand extension—to Facebook as Diet Coke is to Coke. Lite is intended for people who use a mobile phone or do not have broadband Internet access or for some other reason need a smaller, less data-intensive window into Facebook that does not consume much bandwidth. It is a stripped-down version of the service without things like videos. As Facebook heads toward 500 million users, Zuckerberg is embracing user segmentation. Facebook has implemented a daunting parade of changes, even as it continues its headlong growth. Zuckerberg was becoming resigned to protests from the relative few so long as more and more people found value in his service. He started saying he couldn't wait for 2010 so he could stop wearing that damned tie.

17

The Future "My goal was never to just create a company."

Mark Zuckerberg is sitting under the beams of an elegant old Swiss restaurant in Davos during the January 2009 World Economic Forum, the celebrated annual gathering of government and industry leaders. To his right is Sheryl Sandberg, and at the other end of the small table is Larry Page, Google's co-founder. Accel Partners, Facebook's primary venture capital investor, is hosting an annual Davos gathering for technologists and scientists called the "Nerd's Dinner." This year Accel has flown in not one but two American sommeliers to present several varieties of $600-a-bottle California wine. Zuckerberg, who has drunk a couple of glasses, leans forward.

"Larry, do you use Facebook?" he asks.

"No, not really," Page replies without affect in his high-pitched nasal voice. Zuckerberg seems disappointed.

"Why not?" he persists.

"It's not really designed for me," Page answers. Zuckerberg starts to ask him another question, but is deterred by Sandberg.

"Mark! Don't talk about that in front of David!" she scolds. (That is me, sitting on Zuckerberg's left.) Sandberg is well-versed in managing journalists.

But in asking such a question so openly of the co-founder of Google, Silicon Valley's king and Facebook's rival in many ways, Zuckerberg displays a few facets of his character. He may be a tad naïve, but he is simultaneously fearless, competitive, and supremely confident, even cocky. He is not afraid of Google, though he remains a little obsessed with it. He really wants Page to like Facebook, but he also wants to see what will happen when he asks.

• • •

Zuckerberg will almost certainly continue to rule over Facebook with absolute authority. He wants to rule not only Facebook, but in some sense the evolving communications infrastructure of the planet. Yet he believes that Facebook's continued success depends on its ability to retain the confidence of its users. As he told users during the vote on the terms of service, he wants to govern Facebook fairly, through an "open and transparent" dialogue. It remains more important to the young CEO to further the honest transparency he believes in and to facilitate more sharing and communicating than to turn Facebook into a profitable business, though he believes he can pursue both goals in tandem.

I once asked Zuckerberg if he ever worried that Facebook could get into financial jeopardy. "Well, there are different levels of jeopardy," he replied. "Is it sustainable? Will it go out of business? I don't spend any time worrying about that. It's fine. Can it be a $10-billion company or something like that? Okay, I think we have a really good chance of getting there."

Some colleagues say Zuckerberg's desire to prioritize openness and fairness over profit shows he is good at delaying gratification. Or maybe he's just so driven that gratification is irrelevant. "He's always striving to do the next thing," says an executive who has worked closely with him. "For most people there are plateaus and milestones you hit and it allows you to sit back and celebrate and feel a sense of accomplishment. That doesn't really exist for Mark."

Zuckerberg's pursuit of growth over money does not seem to have diminished Facebook's financial prospects. Board member Marc Andreessen has as much perspective on these matters as anyone. "Mark has never argued that Facebook will not make a lot of money," observes Andreessen. "On the financial side really it's a timing thing. Focusing on anything other than establishing a global franchise is a waste of time."

Andreessen is among the very small number of people Zuckerberg regularly turns to for advice, so his views count. ("Marc is in a position to say things and have Mark Z. believe it. I don't think any of the rest of us are," says David Sze of Greylock Partners, a big Facebook investor.)

Andreessen's strong advice is to keep investing for growth. He explains himself in a fall 2009 interview in a cushy Silicon Valley hotel lobby, talking so fast it's lucky I have a recorder. "How much cash has the company burned to date?" he asks. "A few hundred million, right? So how many active users do they have? Three hundred million? So the company has spent a dollar or so per active user and has built a global franchise, a global brand, with real staying power, stickiness, network effects, R&D, competitive advantage, and a whole future roadmap of technology that's on its way out the door. For a dollar a user? Like, you would do that over and over and over again.

"So, okay, let's ask the question—What if the potential here is to get to 500 million active users, or a billion active users, or two billion, right? Would you keep spending that dollar to get there? Of course! The answer is—of course! You would. Compare that to the cost of building anything of any similar scale and you'd say you have the bargain of the century." Andreessen is very tall, and he leans his large, bulbous, shaven head in toward me as his forceful words careen toward what is in his view an irrefutable conclusion. He is difficult to argue with. If you have him on your board he will carry a lot of sway. No matter. He and Zuckerberg agree.

Zuckerberg's mentors and advisors have evolved as the company has grown, from Eduardo Saverin, his friend who knew something about business, to Sean Parker, who had started companies and knew how to deal with financiers, to Don Graham, who ran one of the country's largest media companies, and now to Andreessen and Steve Jobs, widely considered the most influential businessman in the world. Zuckerberg admires Jobs and has been spending an increasing amount of time with him.

Facebook's board has always been small. Thanks to the machinations of Sean Parker in 2004, Zuckerberg has always controlled it. He expects it to support him in his long-term approach to managing the company. When I ask Andreessen what he thinks about Zuckerberg's control of the company, he blurts out, "Oh, that's a good thing." Only very strong founder CEOs, he says, can build big enduring tech companies. He compares Zuckerberg to Bill Gates, Jeff Bezos, and Jobs himself.

Each board member works with Zuckerberg in his own way. Jim Breyer, who joined when Accel invested in 2005, weighs in on organizational structure and hiring. ("Mark always wanted a hacker culture and creative chaos," says Breyer. "My point to him is you want that around product innovation but not in areas like sales, human resources, or legal.") Andreessen gets involved in management but also with product design. He feels protective of Zuckerberg, and tries to keep him from making the same mistakes he made as a young entrepreneur. On the other hand, Peter Thiel, appointed when he invested $500,000 in 2004, is less interested in management and talks to Zuckerberg more about long-term corporate strategy and the overall economic environment. Zuckerberg describes their ongoing discussions: "It's mostly like 'Raise money now.' 'Don't raise money.' 'Keep this money in the bank.' 'You should sell the company now.' 'You should not sell the company now.' I listen to him on that."

Zuckerberg talked about adding Don Graham of the Washington Post Company to the board as far back as 2005, even after Accel outbid Graham to invest in Thefacebook. But both agreed then that the company was still too small. Zuckerberg landed Graham in 2009, finally filling all five board seats (though both Graham and Andreessen serve at his pleasure). Zuckerberg admires Graham's long-term view of his business as well as the structure of the Washington Post Company that enables it.

In November 2009, Zuckerberg implemented a shareholder arrangement at Facebook similar to that of the Washington Post Company. It assures that he and his allies—all existing shareholders were converted to the new "Class B" stock with the change—will retain control of Facebook after it goes public. Google had created such a structure for its own IPO in August 2004. Afterward, management and directors controlled 61 percent of Google's voting power through shares that carried ten votes each, while common shares were allotted one vote. Facebook's new share structure has identical voting provisions. Facebook won't go public until it reaches at least $1 billion in annual sales, a level it will almost certainly achieve in 2010. But board member Jim Breyer said in January 2010 that the company would definitely not go public this year. When the company goes public will depend on whether Zuckerberg believes that an IPO will benefit the company in other ways, for example by making it more prominent in the ranks of business. He will never do it just because he wants to cash out his own wealth. And once he does take it public, he will face inevitable pressures from Wall Street. It will become considerably harder to maintain his resolute emphasis on his vision of sharing and on growth above short-term revenue.

Zuckerberg owns about 24 percent of Facebook's stock, worth about $3 billion at the price the stock traded at privately as of early 2010. The second largest block is Accel's at about 10 percent, plus about 1 percent controlled by Jim Breyer personally (the result of that $1 million he invested back in 2005). Dustin Moskovitz owns about 6 percent. In may 2009 Russia's Digital Sky Technologies bought 2 percent from the company, about 1.5 percent from miscellaneous employees, and later another 1.5 percent from various holders to make it Facebook's second-largest outside investor, with about 5 percent. Eduardo Saverin owns 5 percent, Sean Parker about 4 percent, and Peter Thiel around 3 percent (he sold about half his holdings in late 2009, mostly to Digital Sky). Greylock Partners and Meritech Capital Partners each have between 1 percent and 2 percent. Microsoft owns about 1.3 percent, and Hong Kong billionaire Li Ka-shing about .75 percent. Advertising giant The Interpublic Group owns a little less than half a percent, the happy heritage of an ad-and-equity deal in Facebook's early days. A amall group of current and former employees own a substantial share but less than 1 percent. They include Matt Cohler, Jeff Rothschild, Adam D'Angelo, Chris Hughes, and Owen Van Natta. Others with sizable holdings include Reid Hoffman and Mark Pincus, who invested alongside Peter Thiel in the company's very first financing round, as well as Western Technology Investments (WTI) which loaned the company a total of $3.6 million in its first two years, and invested $25,000 in that same initial round. Employees and outside investors own the remaining 30 percent or so.

Though Digital Sky bought its shares from the company at a price that valued it at $10 billion, it's hard to say what Facebook is really worth. As recently as late 2008 the so-called fair market value placed on its shares gave it a total value of only about $2.5 billion. This is the price Accel, for instance, placed then on its own Facebook shares for bookkeeping purposes. "I know it will end up a big number someday," Breyer told me around that time. "So I don't really care what it is now." (His Accel VC firm purchased some of the employee shares along with Digital Sky at a valuation of about $7.5 billion in mid-2009.) Breyer's fellow board member Thiel, however, is not so certain. "The range of what it may be worth from here is extremely big," he said in an early 2009 interview. "It may be worth a lot more. It may be worth nothing at all." These guys must have pretty interesting conversations in the boardroom. Thiel also talks about "the incredibly high anxiety levels people have about—is this going to be the most successful thing ever, or is it going to weirdly spiral out?" Even though the company remains private, insiders have begun periodically selling Facebook's stock on specialty exchanges like SecondMarket and SharesPost, at prices that put the company's valuation as high as $14 billion by early 2010.

For all his conviction about the inevitability of ever-growing transparency, Zuckerberg remains concerned about a corollary issue—who controls your information. "The world moving towards more transparency could be the trend driving the most change over the next ten or twenty years," he says, "assuming there's no massive act of violence or other political disruption. But there's still a big question about how that happens. When you ask people what they think about transparency, some get a negative picture in their mind—the vision of a surveillance world. You can paint some really dystopian futures. Will the transparency be used to centralize power or to decentralize it? I'm convinced that the trend towards greater transparency is inevitable. But I honestly don't know how this other piece [whether or not we'll be subject to constant surveillance] plays out.

"Let me paint the two scenarios for you. They correspond to two companies in the Valley. It's not completely this extreme, but they are on different sides of the spectrum. On the one hand you have Google, which primarily gets information by tracking stuff that's going on. They call it crawling. They crawl the Web and get information and bring it into their systems. They want to build maps, so they send around vans which literally go and take pictures of your home for their Street View system. And the way they collect and build profiles on people to do advertising is by tracking where you go on the Web, through cookies with DoubleClick and AdSense. That's how they build a profile about what you're interested in. Google is a great company . . ." He hesitates. "But you can see that taken to a logical extreme that is a little scary.

"On the other hand, we started the company saying there should be another way. If you allow people to share what they want and give them good tools to control what they're sharing, you can get even more information shared. But think of all the things you share on Facebook that you wouldn't want to share with everyone, right? You wouldn't want these things to be crawled or indexed—like pictures from family vacations, your phone number, anything that happens on an intranet inside a company, or any kind of private message or email. So a lot of stuff is getting more and more open, but there's a lot of stuff that's not open to everyone.

"This is one of the most important problems for the next ten or twenty years. Given that the world is moving towards more sharing of information, making sure that it happens in a bottom-up way, with people inputting the information themselves and having control over how their information interacts with the system, as opposed to a centralized way, through it being tracked in some surveillance system. I think that's critical for the world." He laughs a little nervous laugh, realizing he sounds awfully passionate. "That's just a really important part of my personality, and what I care about."

Facebook's position in all this is not entirely benign, despite Zuckerberg's high-minded tone. Regardless of Google, Facebook has not always protected personal information carefully. It initially made wrong choices about such information in the News Feed, Beacon, and terms-of-service incidents. It got enormous criticism in the late 2009 for strongly encouraging members to select the "everyone" privacy settings for their personal information.

For all the protections it can offer our data from the potential depradations of others, Facebook the company will always be able to see our data. It is itself a centralizer, gathering all this information about us under one corporate umbrella. It is comforting that Zuckerberg is so personally passionate about the importance of protecting people from information predators. But what guarantee could Facebook's users possibly get that his good intentions will last indefinitely? In a worst-case scenario, possibly in some future when Zuckerberg has lost control of his creation, Facebook itself could become a giant surveillance system.

Zuckerberg's big-picture adviser and board member Thiel makes a similar point about Google. It's clearly something they've spent time talking about. "Google in many ways is an incredible company with an incredible founding vision," says Thiel. "But a very profound difference is, I think, at its core Google believes that at the end of this globalization process the world will be centered on computers, and computers will be doing everything. That is probably one of the reasons Google has missed the boat on the social networking phenomenon. I don't want to denigrate Google. The Google model is that information, organizing the world's information, is the most important thing.

"The Facebook model is radically different. One of the things that is critical about good globalization in my mind is that in some sense humans maintain mastery over technology, rather than the other way around. The value of the company economically, politically, culturally—whatever—stems from the idea that people are the most important thing. Helping the world's people self-organize is the most important thing."

Some aspects of the contrast Zuckerberg and Thiel point to is already evident. Facebook poses a concrete threat to Google's mandate to index and organize the world's information. "What happens on Facebook's servers stays on Facebook's servers," wrote Fred Vogelstein in an insightful July 2009 article in *Wired* magazine titled "The Great Wall of Facebook." "That represents a massive and fast-growing blind spot for Google." Insiders at the search company confirm that this is a much-discussed worry there. If data inside the largest and fastest-growing Web service is off-limits to Google, its ability to serve as the definitive search site could be in jeopardy. The quantity of information we're talking about is considerable. Status updates alone on Facebook are estimated by company insiders to amount to more than ten times more words than on all blogs worldwide. The Compete research firm reports that in January 2010, 11.6 percent of all online time in the United States was spent on Facebook, vs. 4.1 for Google.

The problem for Google is compounded as personal information begins to help us all search for information online. If a friend has previously found benefit from some data source, or purchased an item you're looking at, that's something you're going to want to know when you conduct a search. In a rare public admission, a Google product manager conceded to journalists at a May 2009 Tokyo meeting that for many types of searches, users find information more trustworthy if it comes from friends, and that Facebook can potentially help users achieve that level of trustworthiness better. Then, in a public appearance in late 2009, Google CEO Eric Schmidt conceded that one of the top challenges his company faces is figuring out how to search, index, and present social media content like that created in Facebook. He called this issue "the great challenge of the age."

Facebook itself continues to improve its own tools for searching content on its site, but it isn't very good at it, either. Now it's possible to search all Facebook commercial pages as well as data for which individuals have removed privacy controls so "everyone" has access. The company aims to further encourage use of the "everyone" setting even as it improves its search tools. This not only tweaks Google, but also helps fend off Twitter, whose success has been aided by the fact that tweets can be searched easily. For standard Internet searching from inside the service, Facebook adds insult to injury by using the Bing search engine from Microsoft, Google's archrival.

The competition between Facebook and Google will remain heated, though it could be resolved a number of ways. One cannot rule out the possibility of a rapprochement—even possibly some sort of deal or business combination that enabled the two companies' data to somehow intermingle, despite Zuckerberg and Thiel's protestations. Google probably would still like to buy Facebook, but as the search giant encounters more and more regulatory and anti-trust resistance, the chances that it would be allowed to make such a purchase diminish rapidly. Alternatively, if Facebook got closer to Microsoft its rivalry with Google could become more acrimonious. Most likely Facebook will continue to play the two giants against one another, as it did when Microsoft invested.

In the meantime, Facebook and Google battle for online market share and mindshare as well as for executives and engineers. Facebook has become the clear number two Internet company worldwide in numbers of users, behind Google, even as it has surpassed Google and all other sites in the total time its users spend there. As for employees, Zuckerberg's hiring of Sheryl Sandberg, as well as top Google communications executive Elliot Schrage, did not go over well at Google. In January 2008, Zuckerberg rode to Davos on the Google jet, chatting much of the way with Sandberg, just prior to hiring her. Neither of them was offered a ride in 2009. Google got some revenge in 2008 when it lured back one of its other prominent defectors. The talented programmer and manager Ben Ling had been in charge of Facebook's platform for only ten months before he returned to Google. A large number of Facebook's employees are former Googlers.

I tried to get Google CEO Eric Schmidt to respond to Zuckerberg's comments about Google and surveillance. "My preference is not to comment on things others say about Google," he replied diplomatically in an email. "Mark has done a masterful job of navigating a tough set of waters over the last few years, and he is obviously an exceptional leader and strategist, especially given his relatively young age."

Facebook may soon begin to share something else with Google: the perception that it has become too big. Regulators in Europe opened a formal antitrust investigation into Google in early 2010. Microsoft became so powerful that the Department of Justice moved to break it up. Though that effort failed, Facebook's ambitions and potential to exert control over both users and platform partners are at least as great. "Facebook controls its platform more tightly than Microsoft ever did," says one close observer. "Facebook can flip a switch and turn you off. Anybody. Anytime." If Facebook keeps growing and Zuckerberg appears not to abide by his intended path of user consultation and corporate benevolence, such a development could begin to invite scrutiny from the world's antitrust enforcers.

The closer Facebook gets to achieving its vision of providing a universal identity system for everyone on the Internet, the more likely it is to attract government attention. Facebook could have more data about citizens than governments do. Canada's national privacy commissioner spent a year examining Facebook privacy policies before negotiating a number of changes announced in August 2009. It was telling that the inquiry emerged from Canada. A larger percentage of Canada's online population is on Facebook—42 percent—than in any other major developed country, according to *The Facebook Global Monitor*.

In the extreme view, Facebook could take over key functions from governments. Says Yuri Milner, the company's big Russian investor: "Facebook Connect is basically your passport—your online passport. The government issues passports. Now you have somebody else worldwide who is issuing passports for people. That is competitive, there's no doubt about it. But who says issuing passports is government's job? This will be global citizenship."

Privacy and identity experts are certain such a transition cannot happen smoothly. Says John Clippinger, an official at the Berkman Center for Internet & Society at Harvard and author of *A Crowd of One: The Future of Individual Identity*: "Facebook is coming up against a crucial civic and legal and national security infrastructure. The identity system is a building block of our civil liberties. Sure, creating social graphs can be a new way of authenticating people. But should Facebook own it? And have no restrictions over it? It is an Orwellian power play. Facebook is trying to control a very fundamental resource and right."

Such views could suggest a rocky road ahead for Facebook. One thing is clear: if Zuckerberg is going to play the role of statesman, he is going to have to spend a lot of time in coming years explaining himself.

As Facebook approaches a membership number in the billions, the need to successfully navigate the shoals of regulation will undoubtedly become a more pressing concern. I asked Thiel about the risk of government intervention. "Facebook will have the maximum amount of legal and political leeway in a world where it's seen as friendly and not threatening," he replied. "I think it isn't threatening, It's not really displacing anybody. I see it as a very hopeful sign that the company has made as much progress as it has, and has received as little resistance as it has. We're at 175 million people [this was February 2009] and no lobbyists in Congress are arguing for Facebook to be shut down."

True enough. But the scrutiny is certainly increasing. For instance, John Borthwick, a prominent New York–based technology investor (he owns a piece of Twitter, among other companies), thinks that in late 2008 Facebook deliberately reset controls that determine whether users receive email notifications of new activity inside the service. Facebook says the resetting was accidental, the result of a technical glitch. But in the process, Borthwick notes, it was able to resume sending all users email notifications about things like messages they'd received. Though he has no proof, he thinks it was a deliberate effort to draw people back into the service in order to increase activity and page views.

Some things Facebook plans will almost certainly create strong outside reaction. For instance, Facebook "credits" could begin to function as a virtual currency, and a transnational one at that. "The currency becomes a way to monetize connections between users," says Facebook's Dan Rose, responsible for monetization. People could use it to transfer money between one another. Since this new mechanism for purchases is identity-based, it might help reduce credit card fraud. And it could enable new conveniences. For instance, you could buy a friend a present online without knowing their address. Just select the present and tell the retailer the name of the friend. The two companies' systems could work out the rest for you, with payment drawn from your Facebook credits. A universal online payment system for hundreds of millions of consumers worldwide might be a huge convenience. It could also trump national boundaries and enable Facebook to begin operating as a truly global economy. But don't be surprised when banks and others begin asking if that should be a role for Facebook.

Zuckerberg professes a deep desire to make sure that Facebook remains a benign force on the Net and in society. "You need to be good in order to get people's trust," he says. "In the past people just didn't expect goodness from companies. I think that's changing now."

"I often say inside the company that my goal was never to just create a company," Zuckerberg explains, staring at me intently as we sit alone in a conference room. "A lot of people misinterpret that, as if I don't care about revenue or profit or any of those things. But what not being just a company means to me is not being just that—building something that actually makes a really big change in the world." His stare is a bit unnerving, but he is concentrating. He continues.

"The question I ask myself like almost every day is 'Am I doing the most important thing I could be doing?'" he says with uncharacteristic expansiveness. "Because if not, we've built this company to a good enough point that I don't have to be doing this, or anything else. That's the argument a lot of people have given me for why we should have sold the company in the past. Then we could just go hang out. So then you face this question of what's important to you. Unless I feel like I'm working on *the most*"—he lingers on these words for emphasis—"important problem that I can help with, then I'm not going to feel good about how I'm spending my time. And that's what this company is."

In the end Mark Zuckerberg's vision is to empower the individual. For him, the most important thing that Facebook can do is to give people tools that enable them to more efficiently communicate and thrive in a world in which more and more information surrounds us all no matter what we do. He wants to help keep individuals from being overwhelmed as large institutions both in business and government gain ever more vast computational and information resources.

His subordinates have mostly come to endorse this way of thinking. "What is the key reason we are at this point, with all this success?" asks Kevin Colleran, Facebook's longest-serving advertising sales executive and a good friend of Zuckerberg. "The key reason is that Mark is not motivated by money." Chris Cox, the vice president of product and who works alongside Zuckerberg almost daily, says, "Mark would rather see our business fail in an attempt to do what is right and to do something great and meaningful, than be a big, lame company." A watchword over the years at Facebook has been "Don't be lame." Cox says it means don't do something just to make more money or because everybody is telling you to. It is Facebook's counterpoint to Google's motto 'Don't be evil.'"

Though Facebook is filling out with executives of all ages, people in their twenties still constitute a critical mass. They understand how Zuckerberg thinks because they are much like him. They take the impact of their work with profound seriousness, even as they seem to spend much of the day wiggling erratically around the vast office on two-wheeled RipStick skateboards. Many naturally gravitated to Facebook after developing deep convictions about the social implications of a service they used daily. When I'm in their offices I often feel this could be the smartest bunch of young people on the planet today. The average age of the 1,400 employees is thirty-one.

The company moved in May 2009 from a hodge-podge of rented offices scattered across downtown Palo Alto into a big 135,000-square-foot former manufacturing plant a few miles across town. The office was deliberately picked for its funky undecorated quality—Zuckerberg and Sandberg didn't want to move into fancy digs like the ones occupied by Google or Yahoo. They talked about the perils of the "you have arrived" corporate office. Their view was that it could lead employees to become complacent. But even the new office was quickly filled and the company rented another even bigger industrial building nearby for further expansion.

Facebook has shown a peculiar durability. Ever since it began, critics have predicted it was at risk of losing its "coolness" and would soon begin to decline: "If it lets in Harvard staff . . . If it goes beyond Harvard . . . If it includes colleges outside the Ivy League . . . If high school kids can join . . . If adults are allowed . . . everyone else will leave." The "end of Facebook" article has become a cliché.

Meanwhile, the service has just kept growing and has not measurably lost the collective allegiance of any class, age, or nationality of user. This trend cannot last forever, but it has shown no signs of stopping yet, as Facebook's relentless internationalization continues. "Even we are still trying to comprehend the scale and power of what it is we've built," marvels Chamath Palihapitiya, vice president for growth and internationalization. "We think this is a company that will build value for decades and decades and decades."

Facebook is changing our notion of community, both at the neighborhood level and the planetary one. It may help us to move back toward a kind of intimacy that the ever-quickening pace of modern life has drawn us away from.

At the same time, Facebook's global scale, combined with the quantity of personal information its users entrust to it, suggests a movement toward a form of universal connectivity that is truly new in human society. The social philosopher and media theorist Marshall McLuhan is a favorite at the company. He coined the term "the global village." In his influential 1964 book, *Understanding Media: The Extensions of Man*, he predicted the development of a universal communications platform that would unite the planet. "Rapidly, we approach the final phase of the extensions of man—the technological simulation of consciousness," he wrote, "when the creative process of knowing will be collectively and corporately extended to the whole of society." We are not there yet. Facebook is not what he describes. The world remains fragmented. But no previous tool has ever extended a "creative process of knowing" so widely.

The overall contributions of Facebook's users constitute a global aggregation of ideas and feelings. Some have gone so far as to say it could evolve toward a kind of crude global brain. The reason people sometimes talk like that is that once all this personal data exists in one place it can be examined by sophisticated software in order to learn new things about aggregate sentiment or ideas. One company project announced in late 2009 is the Gross National Happiness Index. Analytic software measures the occurrence over time of words and phrases on Facebook that suggest happiness or unhappiness. That produces a chart that is intended to be "indicative of how we are collectively feeling," according to a post on the Facebook Blog. Initially it will only plumb data produced by English-speakers in the United States. But over time it will likely be extended more broadly, creating an unprecedented gauge of global sentiment. Such tools will become steadily more capable.

Facebook embodies stunningly efficient qualities of universal connectivity. Go to its search box and type in the name of anyone you've ever met. The chances are good you will be directed to a page with their name and photo. If you want, from there you can send them a message. Facebook aims to assemble a directory of the entire human race, or at least those parts of it that are connected to the Internet. It creates a direct pathway between any two individuals.

These capabilities might conceivably over time lead to more global understanding. Or perhaps they won't. Maybe we will use Facebook merely to connect more closely to those we already know. Maybe doing that will reinforce our sense of tribal separation.

After all, Zuckerberg's original conception of Facebook, maintained rigorously until recently, was of a service to communicate with people you already are acquainted with in real life. As Facebook encountered the need to build its revenues, it embraced commercial pages and a marketing culture that has come to coexist alongside a culture of personal interconnection. Then, as the Twitter challenge emerged, it expanded its self-definition further to become a service where people communicate with everyone as well as with their friends. In some ways this was a natural consequence of another of Zuckerberg's founding premises—that "sharing" and transparency were becoming irresistible elements of modern experience.

But reciprocal personal connections packed with very private data may not coexist well with unbridled sharing. Does it really make sense to combine the original conception of Facebook with what Twitter and MySpace and a host of other, less-restrictive services do? Can a system based on trust ever evolve to become truly open?

The answer to such questions will depend on decisions that Facebook makes as it refines and improves its service. Zuckerberg cares deeply about Facebook's potential to serve as a bridge between people. He will work to turn it ever more into a town square for the global village. But as we have heard, he also has a conviction about the importance of helping people protect their most sensitive personal data. Maintaining the enthusiasm of hundreds of millions of people who joined originally to communicate with their friends will remain his ongoing challenge.

By the time you read this book, Facebook will likely have exceeded half a billion active users. It announced 400 million in February 2010 but has been adding about 25 million new users every month.

The company is increasingly embedded in the fabric of modern life and culture. One frequently overhears the word "Facebook" in conversations in public places in almost every country, no matter what the language. One dictionary named "unfriend" the 2009 word of the year. TV shows embed Facebook in their story line.

Facebook's social impact continues to broaden. For many adults all over the world, it has rekindled moribund relationships. Jon Weisblatt, a marketing consultant in Austin, Texas, wrote a note on the Facebook page I maintain devoted to this book (www.facebook.com/thefacebookeffect) in which he coined the phrase "Facebook vertigo." That's the feeling he gets "when I suddenly see the names and faces of friends from long ago." And for those looking for love, Facebook represents a chance to try again. Since more or less every former friend is accessible via a simple Facebook message, many are tempted to get back in touch with crushes from high school or college. So many people have renewed relationships in this way that a word has emerged to describe them—"retrosexuals."

But Facebook has also become yet another place for the antisocial to wreak havoc. Vandals and commercial miscreants now frequently set up phony sites that look like Facebook in order to harvest people's passwords. Then they log in to Facebook using the stolen password and send spam messages to that person's friends, often with the aim of stealing yet more passwords. Such a "phishing" page fooled even the chairman of the U.S. Federal Communications Commission, Julius Genachowski. A bunch of his Facebook friends received a cryptic message reading, "Adam got me started making money with this."

Facebook's competitors in general-interest social networking are struggling. MySpace is losing money. Bebo, acquired in 2008 by AOL for $850 million, is now for sale. Both MySpace and Classmates.com, a long-standing if rudimentary social network focused on connecting old high school friends, have put some of their own services onto Facebook's platform and begun to use Facebook Connect.

For Zuckerberg one uncomfortable price of success is celebrity. He cannot dine out in Silicon Valley without being recognized and interrupted by people who ask for his autograph or take his picture.

A friend of mine lives in Palo Alto a few blocks from Facebook's offices. One weekend he was returning late at night from a long day of family activities with a car full of irritable children. He and his wife were relieved to finally be home. But as he approached his driveway, his car's headlights silhouetted a man standing on the sidewalk, blocking his path.

The small man with curly hair didn't notice them. He was oblivious, immobilized, hands clasped behind his back, head down, lost in thought. There was a gravity in the man's demeanor. My friend paused. Despite his family's exhaustion, his instinct told him not to interrupt. He waited. After a minute or so, the pensive Mark Zuckerberg looked up and continued slowly down the sidewalk.

Acknowledgments

Thanks go first to Mark Zuckerberg. Had he not encouraged me to write this book and cooperated as I did so, it would likely not have happened. As I proceeded, I often said to myself and to others how much I liked writing a book about someone so committed to transparency. He tried hard to answer even questions that had embarrassing answers.

It would have been impossible to spend so much time on this project without the support and love of my wife, Elena Sisto, and my daughter, Clara Kirkpatrick, who also often served as a two-person Facebook focus group. They formed the core team.

At Simon & Schuster I was blessed with not just one but two terrific editors. Bob Bender ran the project and oversaw everything with the aplomb and judgment that makes him such an admired industry veteran. In addition, Dedi Felman's advice on structure and her surgical editing were invaluable.

My superb agent Wayne Kabak guided me throughout the process with sage counsel, and I am deeply grateful. Thanks also to Jim Wiatt for convincing me I should write a book. Teri Tobias is my terrific international agent.

Julia Lieblich was a partner. Without her help I couldn't have pulled it off. Judy Adler was another critical ally.

At Facebook, Brandee Barker was my guru. She spent innumerable hours helping me figure out who to talk to and sitting patiently as I did so. Elliot Schrage, who heads all communications for the company, was hugely supportive and helpful. Larry Yu also did yeoman's service in the interview process, and Maureen O'Hara performed frequent scheduling miracles.

My close friend Brent Schlender read proposals and weighed in throughout with advice honed by a quarter century covering technology. Jessi Hempel contributed in a variety of ways. Other friends who helped include Jim Aley, Marc Benioff, Lynne Benioff, Brett Fromson, Frank Levy, Ellen McGirt, Rick Moody, Peter Petre, Julie Schlosser, and Della Van Heyst. Justin Smith and John Battelle contributed wisdom from the trenches. Tedd Ross Pitts and Gretl Rasmussen worked on the hard stuff. Ali Axon cheered me on.

Special thanks also to Matt Cohler, Joe Green, Chris Hughes, Dustin Moskovitz, and Sean Parker.

Facebook cooperated extensively in the preparation of *The Facebook Effect*, as did CEO Mark Zuckerberg. Almost nobody connected to the company refused to talk to me. However, there was no quid pro quo. Facebook neither requested nor received any rights of approval, and as far as I know, its executives did not see the book before it went to press. Company employees, when confronted with a particularly probing question, periodically stopped and turned quizzically to the Facebook public relations person who was often nearby, but they were without exception encouraged to answer my question. And I talked to many people without supervision.

Some people submitted to multiple interviews. First among these is Mark Zuckerberg himself. Others who were especially generous with their time included Jim Breyer, Matt Cohler, Chris Cox, Kevin Efrusy, Joe Green, Chris Hughes, Chris Kelly, Dave Morin, Dustin Moskovitz, Chamath Palihapitiya, Sean Parker, Dan Rose, Sheryl Sandberg, and Aaron Sittig.

Other interviews at Facebook included Carolyn Abram, Aditya Agarwal, Ethan Beard, Charlie Cheever, Kevin Colleran, Adam D'Angelo, Gareth Davis, Dave Fetterman, Anikka Fragodt, Naomi Gleit, Jonathan Heiliger, Matt Jacobson, Meagan Marks, Scott Marlette, Cameron Marlow, Mike Murphy, Javier Olivan, Jeff Rothschild, Ruchi Sanghvi, Barry Schnitt, Mike Schroepfer, Peter Thiel, Gideon Yu, and Randi Zuckerberg.

I spoke to many people who have interacted with or closely observed Facebook over its short history. In addition to some who prefer to remain unnamed, I interviewed Jonathan Abrams, Marko Ahtissari, Saeed Amidi, Marc Andreessen, Tim Armstrong, Samir Arora, Kevin Barenblat, Hank Barry, Tom Bedecarre, Gina Bianchini, Tricia Black, Rene Bonvanie, Jeremy Burton, Michele Clarke, Jared Cohen, Ron Conway, John Clippinger, Tom Crampton, Sebastian de Halleux, Soumitra Dutta, Nick Earle, Dani Essindi, Rahim Fazal, Lukasz Gadowski, Bill Gates, Seth Goldstein, Susan Gordon, Don Graham, Robert Hertzberg, Doug Hirsch, Reid Hoffman, Ken Howery, Joshua Iverson, Karl Jacob, Rebecca Jacoby, Bruce Jaffe, Josh James, Jeff Jarvis, Suzanne McGee, Mike Lazerow, Tara Lemmey, Sam Lessin, Max Levchin, Titus Levy, Charlene Li, Caroline Little, Chris Ma, Olivia Ma, Marissa Mayer, Oscar Morales, Yuri Milner, Rick Murray, Mairtini niDhomhnaill, Ray Ozzie, Philipp Pieper, Mark Pincus, Shervin Pishevar, Jeff Pulver, Scott Rafer, J. P. Rangaswami, Andrew Rasiej, Robin Reed, Gerry Rosberg, John Rosenthal, Marc Rotenberg, Geoff Sands, Marc Schiller, David Schlesinger, Clara Shih, Anu Shukla, Megan Smith, Justin Smith, Gary Spangler, Stan Stalnaker, Daniel Stauffacher, Seth Sternberg, Nick Summers, David Sze, Don Tapscott, Rodrigo Teijeiro, Owen Van Natta, Erik Wachtmeister, Duncan Watts, Bill Weaver, Andrew Weinreich, Maurice Werdegar, John Winsor, Michael Wolf, and Robert Wright.

My sincere thanks to them all. I've tried to get it right.

Notes

Prologue: The Facebook Effect

Page

1   *Then abruptly in late December the guerrillas announced*: Ian James, "Venezuelan Mission Heads to Colombia," *Boston Globe*, December 29, 2007, http://www.boston.com/news/world/latinamerica/articles/2007/12/29/venezuelan_mission_heads_to_colombia/ (accessed November 15, 2009).

5   *Had she been personally injured by FARC*: Juliana Rincon Parra, "Colombia: United in a March Against the FARC," *Global Voices Online*, February 5, 2008, http://globalvoicesonline.org/2008/02/05/colombia-the-world-united-in-a-multitudinary-march/ (accessed November 15, 2009).

8   *In mid-2008 a Facebook group organized a huge water fight*: Katie Franklin, "Facebook Water Fight Ruins Prized Garden," *Daily Telegraph*, May 7, 2008, http://www.telegraph.co.uk/news/newstopics/howaboutthat/1935926/Facebook-water-fight-ruins-prized-garden-in-Leeds.html (accessed November 15, 2009).

9   *They heard about the pillow fight on Facebook*: Angela Cunningham, "Facebook Phenomenon Hits Grand Rapids, Literally!," WZZM-13, September 28, 2008, http://www.wzzm13.com/news/news_story.aspx?storyid=99194&catid=14 (accessed November 15, 2009).

14   *The average Facebook user has*: email from Brandee Barker, Facebook public relations (February 24, 2010).

16   *Now users around the world*: email from Brandee Barker, Facebook public relations (February 24, 2010).

16   *About 108 million Americans are active on Facebook*: Justin Smith, *The Facebook Global Monitor: Tracking Facebook in Global Markets* (Palo Alto, CA: Inside Network, February 2010).

1. The Beginning

Page

21   *From his early years Zuckerberg had*: Sarah Lacy, *Once You're Lucky, Twice You're Good: The Rebirth of Silicon Valley and the Rise of Web 2.0* (New York: Gotham Books, 2008), 141.

23   *the* Harvard Crimson *later called it "guerrilla computing"*: "M*A*S*H—Online 'Facemash' Site, While Mildly Amusing, Catered to the Worst

Side of Harvard Students," *Harvard Crimson*, November 6, 2003, http://www.thecrimson.com/article/2003/11/6/mash-for-the-most-monastic-undergraduates/ (accessed November 15, 2009).

24   *The* Crimson *somewhat eloquently opined*: Ibid.

24   *"We Harvard students could indulge our fondness"*: Ibid.

24   *By the time that happened, around 10:30 P.M.*: Bari Schwartz, "Hot or Not? Website Briefly Judges Looks," *Harvard Crimson*, November 4, 2003.

25   *Zuckerberg was accused of violations*: Katharine Kaplan, "Facemash Creator Survives Ad Board," *Harvard Crimson*, November 19, 2003.

28   *In a December 11 editorial titled*: "Put Online a Happy Face: Electronic Facebook for the Entire College Should Be Both Helpful and Entertaining for All," *Harvard Crimson*, December 11, 2003.

28   *"Much of the trouble surrounding the facemash"*: "M*A*S*H—Online 'Facemash' Site, While Mildly Amusing, Catered to the Worst Side of Harvard Students," *Harvard Crimson*, November 6, 2003.

31   *By Sunday—four days after launch*: Alan J. Tabak, "Hundreds Register for New Facebook Website," *Harvard Crimson*, February 9, 2004.

31   *Zuckerberg later told the* Crimson *that he "hoped"*: Ibid.

33   *"The nature of the site," he told the paper*: Ibid.

33   *Wrote junior Amelia Lester*: Amelia Lester, "Show Your Best Face: Online Social Networks Are a Hop, Click and Jump From Reality," *Harvard Crimson*, February 17, 2004, http://www.thecrimson.harvard.edu/article/2004/2/17/show-your-best-face-lets-talk/ (accessed December 11, 2009).

38   *Zuckerberg and Saverin each agreed*: email from Brandee Barker, Facebook public relations (December 11, 2009): "Mark believes each were going to put $10K or $20K in."

40   *Back in that* Crimson *opinion piece*: Ibid.

2. Palo Alto

Page

46   *In the end they even hired*: Lacy, 147.

47   *Parker had to overdraw*: Lacy, 148.

64   *Zuckerberg and Moskovitz were planning to launch*: Olivia Ma, "Need Help? Check Down the Hall," *Newsweek*, August 2, 2004, *http://www.newsweek.com/id/54735/* (accessed December 11, 2009).

3. Social Networking and the Internet

Page

66   *In a 1968 essay by J. C. R. Licklider:* J. C. R. Licklider and Robert Taylor, "The Computer as a Communication Device," *Science and Technology* (April 1968), http://www.kurzweilai.net/meme/frame.html?main=/articles/art0353.html (accessed December 11, 2009).

67   *"A virtual community is a group of people":* Howard Rheingold, "Virtual Communities—Exchanging Ideas Through Computer Bulletin Boards," *Whole Earth Review* (Winter 1987), http://findarticles.com/p/articles/mi_m1510/is_n57/ai_6203867/ (accessed November 15, 2009).

67   *Two Internet sociologists, danah boyd and Nicole Ellison:* danah boyd and Nicole Ellison, "Social Network Sites: Definition, History, And Scholarship," *Journal of Computer-Mediated Communication* 13 (2007), http://jcmc.indiana.edu/vol13/issue1/boyd.ellison.html (accessed November 15, 2009).

69   *Nonetheless, by 1999 sixdegrees had reached:* Details about sixdegrees from interview and email followup with Andrew Weinreich, 2009.

74   *But, according to* Stealing MySpace*:* Julia Angwin, *Stealing MySpace: The Battle to Control the Most Popular Website in America* (New York: Random House, 2009), 52.

76   *In 2003, Angwin notes, the percentage of Americans:* Ibid.

77   *Buyukkokten himself once bragged:* Luke O'Brien, "Poking Facebook," *02138 Magazine* (November 2007), www.02138mag.com/magazine/article/1724.html (accessed November 28, 2009).

80   *Previously they'd won a gold medal:* Ibid.

83   *The civil lawsuit filed on behalf of the three alleges:* ConnectU, Inc. v. Facebook, Inc. et al., in "Justia News And Commentary," *Justia.com*, http://news.justia.com/cases/featured/massachusetts/madce/1:2007cv10593/108516/ (accessed November 28, 2009).

84   *In September 2004 when they filed suit:* Marcella Bombardieri, "Online adversaries: Rivalry between college-networking websites spawns lawsuit," *Boston Globe* (September 17, 2004), http://www.boston.com/news/education/higher/articles/2004/09/17/ (accessed December 27, 2009).

84   *It gave the ConnectU's creators plenty of money:* Michael Liedtke, "Facebook Appraisal Pegs Company's Value At $3.7B," *San Francisco Chronicle* (February 11, 2009), http://www.sfgate.com/cgi-bin/article.cgi?f=/n/a/2009/02/10/state/n230703s73.dtl (accessed November 28, 2009).

84   *In his autobiography:* Aaron Greenspan, *Authoritas: One Student's Harvard Admissions and the Founding of the Facebook Era* (Palo Alto, CA: Think Press, 2008).

4. Fall 2004

Page

86   *He and his parents had loaned the company:* Kevin J. Feeney, "Business, Casual," *Harvard Crimson,* February 24, 2005, http://www.thecrimson .com/article/2005/2/24/business-casual-a-year-ago-mark/ (accessed November 28, 2009).

89   *Thiel told Zuckerberg:* Lacy, 154.

97   *Achilles: Now you know:* Quotes for Achilles (character) from *Troy* (2004), *IMDB,* http://www.imdb.com/character/ch0004244/quotes (accessed November 15, 2009).

97   *beginning to wend its way through federal court:* Feeney, "Business, Casual."

5. Investors

Page

110   *Investor interest was further heightened:* Rebecca Trounson, "'Hi, What's Your Major?' Is Reinvented on Website," *Los Angeles Times* (January 23, 2005), http://articles.latimes.com/2005/jan/23/local/me-facebook23 (accessed December 27, 2009).

113   *He and Moskovitz would have gotten close to $10 million:* Sean Parker, interview with the author.

115   *By early 2004 Tickle had become:* Interview with Samir Arora, January 11, 2010. (Arora was chairman of the board of Tickle in early 2004.)

6. Becoming a Company

Page

129   *Zuckerberg had to be careful which business card:* Karel M. Baloun, *Inside Facebook: Life, Work and Visions of Greatness* (Bloomington, IN: Trafford, 2007), 23.

129   *The main method initially was a wooden figure of an Italian chef:* Ibid., 38.

137   *"Zuck would come into the office and, seeing every chair full":* Ibid., 18.

138   *Zuckerberg's dry wit and classicism showed:* Ibid., 22.

138   *In the spring, MySpace founders Chris DeWolfe:* Angwin, *Stealing MySpace,* 121.

143   *By year-end, it had $5.7 million left:* ConnectU, Inc. v. Facebook, Inc. et al.

7. Fall 2005

Page

149   *As the school year resumed in the fall of 2005*: Michael Arrington, "85% of College Students Use Facebook," *TechCrunch*, September 7, 2005, www.techcrunch.com/2005/09/07/85-of-college-students-use-facebook/ (accessed November 15, 2009).

151   *One new group was called "You're Still in High School . . . "*: John Cassidy, "Me Media: How Hanging Out On The Internet Became Big Business," *New Yorker*, May 15, 2006, http://www.newyorker.com/archive/2006/05/15/060515fa_fact_cassidy (accessed December 11, 2009).

151   *At the beginning of the school year, Facebook had nearly doubled*: Owen Van Natta, interview with author, May 15, 2007.

152   *Ever vigilant about competitors*: Angwin, *Stealing MySpace*, 140, 177.

153   *Zuckerberg was dismissive*: Ibid., 177.

156   *By early 2010 Facebook was hosting*: email from Brandee Barker, Facebook public relations, February 24, 2010.

8. The CEO

Page

166   *"I want to stress the importance of being young"*: Mark Coker, "Start-Up Advice For Entrepreneurs, From Y Combinator Startup School," *Venturebeat*, March 26, 2007, http://venturebeat.com/2007/03/26/start-up-advice-for-entrepreneurs-from-y-combinator-start-up-school/ (accessed November 28, 2009).

169   *But at the end of March, BusinessWeek's online edition*: Steve Rosenbush, "Facebook's on the Block," *BusinessWeek*, March 28, 2006, http://www.businessweek.com/technology/content/mar2006/tc20060327_215976.htm (accessed November 15, 2009).

170   *But to Zuckerberg, what was more significant*: Ibid.

171   *Another imitator, which launched around the same time in China*: Baloun, *Inside Facebook*, 95.

173   *He also quoted a sociologist who speculated*: Cassidy, "Me Media."

174   *who he had met while*: Lacy, 162.

174   *After some negotiation, Zuckerberg*: Lacy, 162.

176   *A week after the program launched*: Rob Walker, "A For-Credit Course," *New York Times*, September 30, 2007, http://www.nytimes.com/2007/09/30/magazine/30wwlnconsumed-t.html (accessed December 27, 2009).

177   *As part of the deal the ad giant*: email from Brandee Barker, Facebook public relations, December 11, 2009.

9. 2006

Page

184   *Peter Thiel, older but very sympathetic*: Lacy, 165.

186   *Some nights, unable to sleep*: David Kushner, "The Baby Billionaires of Silicon Valley," *Rolling Stone*, November 16, 2006, http://rollingstone .com/news/story/12286036/the_baby_billionaires_of_silicon_valley (accessed November 28, 2009).

186   *"I hope he doesn't sell it"*: Kevin Colleran, interview with the author.

190   *Within about three hours the group's membership*: Tracy Samantha Schmidt, "Inside the Backlash Against Facebook," *Time*, September 6, 2006, www.time.com/time/nation/article/0,8599,1532225,00.html (accessed December 11, 2009).

190   *And there were about five hundred other protest groups*: Brandon Moore, "Student users say new Facebook feed borders on stalking," *Arizona Daily Wildcat*, September 8, 2006, http://wildcat.arizona.edu/2.2257/student-users-say-new-facebook-feed-borders-on-stalking-1.177273 (accessed December 11, 2009).

190   *"Chuck Norris come save us"*: Layla Aslani, "Users Rebel Against Facebook Feature," *Michigan Daily*, September 7, 2006, http://www.michigandaily. com/content/users-rebel-against-facebook-feature (accessed December 11, 2009).

190   *"You shouldn't be forced to have a Web log"*: Moore, "Student Users."

190   *"I'm really creeped out"*: Aslani, "Users Rebel."

191   *But Zuckerberg, in New York on a promotional trip*: Andrew Kessler, "Weekend Interview with Facebook's Mark Zuckerberg," *Wall Street Journal*, March 24, 2007, http://www.andykessler.com/andy_kessler/2007/03/ wsj_weekend_int.html (accessed December 11, 2009).

10. Privacy

Page

200   *As one expert in privacy law recently asked*: James Grimmelmann, "Saving Facebook," *Iowa Law Review* (2009), http://www.law.uiowa.edu/journals/ ilr/Issue%20PDFs/ILR_94–4_Grimmelmann.pdf (accessed December 11, 2009).

201   *"At every turn, it seems Facebook makes it more difficult"*: Marc Rotenberg, "Online Friends At What Price?," *Sacramento Bee*, July 20, 2008.

203   *She defined it as "being able to keep in touch with people"*: Leisa Reichelt, "Ambient Intimacy," *Disambiguity*, March 1, 2007, http://www.disambiguity.com/ambient-intimacy/ (accessed November 15, 2009).

204 *A widely discussed 2008 article in the* New York Times: Clive Thompson, "Brave New World of Digital Intimacy," *New York Times Magazine*, September 7, 2008, http://www.nytimes.com/2008/09/07/magazine/07awareness-t.html (accessed December 11, 2009).

205 *Facebook membership is becoming common:* Catherine Arnst, "Kids on Facebook," *BusinessWeek Working Parents Blog* (January 9, 2010), http://www.businessweek.com/careers/workingparents/blog/archives/2010/01//kids_on_facebook.html (accessed January 17, 2010).

205 *A guard at a Leicester, England, prison:* "Facebook Prison Officer Is Sacked," *BBC News*, March 23, 2009, http://news.bbc.co.uk/go/pr/fr//2/hi/uk_news/england/leicestershire/7959063.stm (accessed November 15, 2009).

205 *A Philadelphia court officer was suspended:* "Facebook Bid Costs Phila. Court Aide," *Philly.com*, April 23, 2009, http://www.philly.com/philly/news/homepage/20090423_Facebook_bid_costs_Phila_court_aide.html (accessed November 15, 2009).

205 *They included images of holidays:* Jon Hemming, "British spy chief's cover blown on Facebook," Reuters, July 6, 2009, http://www.reuters.com/article/idUSTRE56403820090705 (accessed December 11, 2009).

206 *The outcome can even be tragic:* "Man Killed Wife in Facebook Row," *BBC News*, October 17, 2008, http://news.bbc.co.uk/2/hi/uk_news/england/london/7676285.stm (accessed November 15, 2009).

206 *The Internet theorist David Weinberger:* David Weinberger, "Weblog Stat Questions," *Joho the Blog: David Weinberger's Weblog*, December 19, 2001, http://www.hyperorg.com/blogger/archive/2001_12_01_archive.html.

207 *At Amherst Regional High School:* Mary Carey, "Spurned High School Student Commits Facebook Revenge," *Amherst Bulletin*, January 9, 2009, http://www.amherstbulletin.com/story/id/124207/ (accessed December 11, 2009).

207 *Those with red plastic cups were spared:* Lisa Guernsey, "Picture Your Name Here," *New York Times*, July 27, 2008, http://www.nytimes.com/2008/07/27/education/edlife/27facebook-innovation.html (accessed December 11, 2009).

208 *These groups—for work, family, college friends:* "Press Room: Product Overview FAQ," Facebook, http://www.facebook.com/press/faq.php (accessed February 24, 2010).

211 *"I didn't even put the company's name":* "Office Worker Sacked For Branding Work Boring On Facebook," *Daily Telegraph*, February 26, 2009, http://www.telegraph.co.uk/technology/facebook/4838076/Office-worker-sacked-for-branding-work-boring-on-Facebook.html (accessed November 28, 2009).

211 *"Most employers wouldn't dream"*: "Facebook Remark Teenager Is Fired," *BBC News*, February 27, 2009, http://news.bbc.co.uk/go/pr/fr/-/2/hi/uk_news/england/essex/7914415.stm (accessed November 15, 2009).

211 *She then sued the principal in federal court*: "Facebook Postings By Students Sparking Legal Fights," *South Florida Sun-Sentinel*, December 13, 2008, www.sun-sentinel.com/news/schools/sfl-flbfacebook1213sb-dec13,0,3402411.story (accessed November 15, 2009).

212 *"There's a deep, probably irreconcilable tension"*: Grimmelmann, "Saving Facebook."

212 *"peer-to-peer privacy violations"*: Ibid.

212 *In Harrison, New York, a police detective*: Swapna Venugopal Ramaswamy, "Facebook 4 Docked Pay; 2 Demoted," *Journal News*, June 6, 2009, http://m.lohud.com/detail.jsp?key=250959&full=1 (accessed November 28, 2009).

213 *An Australian woman named Elmo Keep*: Asher Moses, "Banned For Keeps On Facebook For Odd Name," *Sydney Morning Herald*, September 25, 2008, http://www.smh.com/au/articles/2008/09/25/1222217399252.html (accessed November 15, 2009).

213 *Others who have had difficulties include Japanese author Hiroko Yoda*: Ibid.

213 *A man in Cardiff, Wales, located a half brother*: Catherine Mary Evans, "The Facebook Reunion: How Two Brothers Got In Touch Again For The First Time In 35 Years," *South Wales Echo*, January 16, 2009, http://www.walesonline.co.uk/news/wales-news/2009/01/16/the-facebook-reunion-how-two-brothers-got-in-touch-again-for-the-first-time-in-35-years-91466–22704778/ (accessed December 11, 2009).

214 *"Here's the major change in the last two years"*: Ben Parr, "Social Media And Privacy: Where Are We Two Years After Facebook News Feed?," *Mashable.com*, September 8, 2008, http://mashable.com/2008/09/08/social-media-privacy-news-feed/ (accessed November 28, 2009).

11. The Platform

Page

216 *Every great technology company goes through*: Thanks to Brent Schlender for these ideas.

222 *I published an article titled*: David Kirkpatrick, "Facebook's Plan to Hook Up the World," *Fortune*, May 29, 2007, http://money.cnn.com/2007/05/24/technology/facebook.fortune/index.htm?postversion=2007052511 (accessed November 28, 2009).

227 *The demographics were already spreading out:* Dustin Moskovitz, interview with author, May 14, 2007.

228 *A couple of young guys in San Francisco:* Nick O'Neill, "Lessons From A Successful Facebook Application Team," *Allfacebook,* June 28, 2007, http://www.allfacebook.com/2007/06/lessons-from-a-successful-facebook-application-team/ (accessed November 28, 2009).

229 *Shortly after Scrabulous launched, Hasbro:* Judith Thurman, "Spreading The Word: The New Scrabble Mania," *New Yorker,* January 19, 2009, http://archives.newyorker.com/?i=2009–01–19#folio=026 (accessed December 11, 2009).

229 *Texas HoldEm had 20.3 million:* App Leadership, Appdata, *Insideface-book,* http://www.appdata.com/leaderboard/apps (accessed December 14, 2009).

230 *The highly complex World of Warcraft:* "World of Warcraft Subscriber Base Reaches 11.5 Million Worldwide," Blizzard Entertainment, press release (December 23, 2008), http://eu.blizzard.com/en-gb/company/press/pressreleases.html?081223

232 *More than 250 of these applications:* emails from Brandee Barker, Facebook public relations (February 24, 2010).

232 *Justin Smith of Inside Facebook estimates:* Michael Learmonth and Abbey Klaassen, "App Revenue Is Poised To Surpass Facebook Revenue," *Advertising Age,* May 18, 2009, http://adage.com/digital/article?article_id=136700 (accessed November 28, 2009).

233 *Numerous Facebook games have revenue:* Ibid.

12. $15 Billion

Page

236 *A little more than a year earlier, its third round:* "Press Room: Facebook Factsheet," Facebook, http://www.facebook.com/press/info.php?factsheet (accessed November 28, 2009).

248 *In fact it was to be her surprise:* Ellen Nakashima, "Feeling Betrayed, Facebook Users Force Site To Honor Their Privacy," *Washington Post,* November 30, 2007, http://www.washingtonpost.com/wp-dyn/content/article/2007/11/29/ar2007112902503.html (accessed November 28, 2009).

249 *Quittner compared twenty-three-year-old Zuckerberg's rash decision making:* Josh Quittner, "RIP Facebook?," *Fortune,* December 4, 2007, http://techland.blogs.fortune.cnn.com/2007/12/04/rip-facebook/ (accessed November 15, 2009).

13. Making Money

Page

265   *Some were as young as fourteen*: Brooks Barnes, "An Animated Film Is Created Through Internet Consensus," *New York Times*, July 16, 2009, http://www.nytimes.com/2009/07/16/movies/16mass.html (accessed December 11, 2009).

265   *Its Facebook page attracted 57,000 members*: Ibid.

272   *Pages had about 5.3 billion fans*: Facebook Press Room: Statistics. http://www.facebook.com/press/info.php?statistics (accessed February 24, 2010).

272   *about twenty million users become new fans*: email communication with Brandee Barker, Facebook public relations department (February 24, 2010).

14. Facebook and the World

279   *when groups there grew to include "Lesbians in Dubai"*: "UAE bans Facebook," *Kipp Report* (February 2008), http://www.kippreport.com/2008/02/uae-bans-facebook/?next=4 (accessed November 15, 2009).

279   *After Italian Facebook groups emerged*: Steve Scherer and Giovanni Salzano, "Facebook Says Italy's Plan To Block Web Content Goes Too Far," Bloomberg, February 12, 2009, http://www.bloomberg.com/apps/news?pid=20601085&sid=a6bncyt8rtlw (accessed November 28, 2009).

279   *Meanwhile, a group called "All Palestinians on Facebook"*: Rory McCarthy, "Israel-Palestine Dispute Moves On To Facebook," *Guardian*, March 20, 2008, http://www.guardian.co.uk/world/2008/mar/20/israelandthepalestinians.facebook (accessed November 28, 2009).

280   *Employees of the city of Naples: Global Faces and Networked Places: A Nielsen Report on Social Networking's New Global Footprint* (New York: Nielsen Company, 2009).

280   *An aide called it a great way to connect*: "Danish PM Jogs With Facebook Fans," *BBC News*, April 18, 2008, http://news.bbc.co.uk/go/pr/fr/-/2/hi/europe/7355434.stm (accessed November 28, 2009).

280   *Obscure Colombian rock bands*: Leila Cobo, "Bands reach fans through Spanish-language Facebook," Reuters, February 22, 2008, http://www.reuters.com/article/idusn2262041620080223 (accessed November 28, 2009).

280   *"I log in three hours a day"*: Nadine El Sayed, "Smile And Say 'Facebook,'" *Egypt Today* (April 2007), www.egypttoday.com/article.aspx?articleid=7293 (accessed November 28, 2009).

281   *"Internationally . . . Facebook is perceived as mainstream": Global Faces and Networked Places.*

284   *A couple of days before I joined him in Madrid:* "Facebook's Priority
      Is Growth Not Profit," *Daily Telegraph*, October 9, 2008, http://www
      .telegraph.co.uk/technology/3358773/Facebooks-priority-is-growth-not-
      profit.html (accessed November 28, 2009).

286   *"The clerics think it is necessary"*: Indra Harsaputra, "Indonesian Clerics
      Want Rules For Facebook," *ABC News*, May 21, 2009, http://abcnews
      .go.com/technology/wirestory?id=7641723 (accessed November 28, 2009).

15. Changing Our Institutions

Page

289   *"You can't ignore 20,000 people"*: Josh Hafenbrack, "Online Political Ac-
      tion Can Spark Offline Change," *South Florida Sun-Sentinel*, April 6,
      2008, http://articles.sun-sentinel.com/2008–04–06/news/0804050256_1_
      facebook-online-networking-new-media (accessed December 11, 2009).

289   *Shortly afterward, the minister of communications:* "Egypt Reverses
      Download-Limit Policy Following Internet Subscribers' Protests," *Daily
      News Egypt*, August 14, 2009.

289   *After tens of thousands joined a Facebook group complaining:* Heru An-
      driyanto and Dessy Sagita, "More Prosecutors Probed In Prita Case,"
      Jakarta Globe, June 9, 2009, http://thejakartaglobe.com/home/more-
      prosecutors-probed-in-prita-case/311188 (accessed November 28, 2009);
      Sherria Ayuandini, "Omni Case: A PR Suicide," *Jakarta Post*, June 5,
      2009, http://www.thejakartapost.com/news/2009/06/05/omni-case-a-pr-
      suicide.html (accessed November 28, 2009).

290   *When police conducted drug raids:* "Maties Unite on Facebook," *News24*,
      November 3, 2008, http://www.news24.com/content/southafrica/news/1
      059/64b4adbddb44457cafea4f83b48c4b49/11–03–2008–03–11/maties_
      unite_on_facebook (accessed November 28, 2009).

290   *Comedian David Letterman made a sexual joke:* Olivia Smith, "Fire
      David Letterman Campaign Takes Root; Protest Planned Over Com-
      ment on Sarah Palin's Daughter," *Daily News*, June 15, 2009, http://
      www.nydailynews.com/entertainment/tv/2009/06/15/2009–06–15_fire_
      david_letterman_campaign_takes_root_protest_planned_over_com-
      ment_on_sarah_p.html (accessed November 28, 2009).

290   *Citizens joined on Facebook to protest a jail expansion:* Michele Clock,
      "City Turns to Twitter, Facebook to Fight Jail Expansion," *San Diego
      Union Tribune*, June 8, 2009, http://www.signonsandiego.com/stories/
      2009/jun/08/1m8social231712-city-turns-twitter-facebook-fight-/ (accessed
      November 28, 2009).

290 *campground for gypsies in Bournemouth, England:* Louise Dunderdale, "Protestors Gather Outside Bournemouth Town Hall," *Bournemouth Daily Echo,* June 24, 2009, http://www.bournemouthecho.co.uk/news/4455424.protestors_gather_outside_bournemouth_town_hall/ (accessed November 28, 2009).

290 *plan by the Philippine House of Representatives:* Joey Alarilla, "Filipinos Tweet, Liveblog House of Representatives Debate on Amending Constitution," *CNET Asia,* June 2, 2009, http://asia.cnet.com/blogs/babelmachine/post.htm?id=63011182 (accessed November 28, 2009); Niña Catherine Calleja, "Outrage vs Constituent Assembly on Facebook," *Philippine Daily Inquirer,* June 9, 2009, http://newsinfo.inquirer.net/topstories/topstories/view/20090609–209641/outrage-vs-constituent-assembly-on-facebook (accessed November 28, 2009).

290 *the relocation to Bermuda of prisoners:* Sarah Titterton, "Life in Paradise As Guantanamo Four Take a Dip, Eat Ice Cream, and Plan First Uighur Restaurant in British Territory of Bermuda," *Daily Mail,* June 15, 2009, http://www.dailymail.co.uk/news/worldnews/article-1192872/guantanamo-4-hit-shops-day-freedom.html (accessed November 28, 2009).

293 *By correlating student membership in Facebook political groups:* Jessica T. Feezell, Meredith Conroy, and Mario Guerrero, "Facebook Is . . . Fostering Political Engagement: A Study Of Online Social Networking Groups And Offline Participation," paper presented at the American Political Science Association Annual Meeting, Toronto, September 3–6, 2009.

294 *"It's the ideal way for her to keep in touch":* Andy Barr, "Palin emerges as Facebook phenom," *Politico,* September 19, 2009, http://www.politico.com/news/stories/0909/27344.html (accessed November 15, 2009).

295 *The commandant of the U.S. Coast Guard updates:* Sagar Meghani, "Pentagon Uses Facebook, Twitter to Spread Message," *U.S. News & World Report,* May 1, 2009, http://www.usnews.com/articles/science/2009/05/01/pentagon-uses-facebook-twitter-to-spread-message.html (accessed November 15, 2009).

295 *Even the Saudi Arabian minister of information:* Faisal J. Abbas, "'Just Add Me': Saudi Information Minister Embraces Social Networking," *Huffington Post,* July 7, 2009, http://www.huffingtonpost.com/faisal-abbas/just-add-me-saudi-informa_b_226281.html (accessed November 15, 2009).

295 *"Zuckerberg . . . realized that Facebook wasn't a tool":* Zachary Stewart, "Five Years of Facebook: How It Redefined What We Consider 'News,'" *Nieman Journalism Lab,* February 4, 2009, http://www.niemanlab.org/2009/02/five-years-of-facebook-how-it-redefined-what-we-consider-news/ (accessed November 15, 2009).

296   *it was via Facebook status updates: BackStory Behind the Scenes,* CNN (January 14, 2010), http://www.cnn.com/video/#/video/world/2010/01/14/bs.haiti.earthquake.cnn?iref=videosearch (accessed January 16, 2010).

16. The Evolution of Facebook

Page

306   *Over 80,000 websites use it:* email from Brandee Barker, Facebook public relations (December 11, 2009).

310   *Jonathan Zittrain, a professor at Harvard Law School:* Jonathan Zittrain, "E Pluribus Facebook," *The Future Of The Internet—And How To Stop It,* April 17, 2009, http://futureoftheinternet.org/e-pluribus-facebook (accessed November 28, 2009).

316   *Facebook's iPhone, BlackBerry, and Google Android applications:* email from Brandee Barker, Facebook public relations (February 24, 2010).

17. The Future

Page

325   *"What happens on Facebook's servers stays":* Fred Vogelstein, "The Great Wall of Facebook: The Social Network's Plan to Dominate the Internet—and Keep Google Out," *Wired,* June 22, 2009, http://www.wired.com/techbiz/it/magazine/17–07/ff_facebookwall (accessed November 15, 2009).

326   *In a rare public admission, a Google product manager:* Jay Alabaster, "Google Increasingly Battles Facebook in Search," *Huffington Post,* May 25, 2009, http://www.huffingtonpost.com/2009/05/25/google-increasingly-battl_n_207449.html (accessed November 15, 2009).

328   *A larger percentage of Canada's online population:* Smith, *The Facebook Global Monitor.*

328   *Says John Clippinger, an official:* John Clippinger, *A Crowd of One: The Future of Individual Identity* (New York: PublicAffairs, 2007).

331   *The average age of the 1,400 employees:* email communication with Brandee Barker, Facebook public relations (February 24, 2010).

Additional Reading

Abelson, H., K. Ledeen, H. Lewis. *Blown to Bits: Your Life, Liberty and Happiness After the Digital Explosion*. Boston: Addison-Wesley, 2008.

Abram, C., L. Pearlman. *Facebook for Dummies*. Hoboken, NJ: Wiley, 2008.

Angwin, Julia. *Stealing MySpace: The Battle to Control the Most Popular Website in America*. New York: Random House, 2009.

Battelle, John. *The Search: How Google and Its Rivals Rewrote the Rules of Business and Transformed Our Culture*. New York: Portfolio, 2005.

Benkler, Yochai. *The Wealth of Networks: How Social Production Transforms Markets and Freedom*. New Haven: Yale University Press, 2006.

Brin, David. *The Transparent Society: Will Technology Force Us to Choose Between Privacy and Freedom?* Reading, MA: Perseus Books, 1998.

Buchanan, Mark. *Nexus: Small Worlds and the Groundbreaking Science of Networks*. New York: W. W. Norton, 2003.

Clippinger, John Henry. *A Crowd of One: The Future of Individual Identity*. New York: PublicAffairs, 2007.

De Jonghe, An. *Social Networks Around the World: How Is Web 2.0 Changing Your Daily Life?* An De Jonghe, 2007.

Falk, Richard. *Religion and Humane Global Governance*. New York: Palgrave, 2001.

Fogg, B. J. *Persuasive Technology: Using Computers to Change What We Think and Do*. San Francisco: Morgan Kaufmann, 2003.

Fraser, M., S. Dutta. *Throwing Sheep in the Boardroom: How Online Social Networking Will Transform Your Life, Work and World*. Hoboken, NJ: Wiley, 2008.

Greenspan, Aaron. *Authoritas: One Student's Harvard Admissions and the Founding of the Facebook Era*. Palo Alto: Think Press, 2008.

Hamel, Gary. *The Future of Management*. Boston: Harvard Business School Press, 2007.

Lacy, Sarah. *Once You're Lucky, Twice You're Good: The Rebirth of Silicon Valley and the Rise of Web 2.0*. New York: Gotham Books, 2008.

Li, C., J. Bernoff. *Groundswell: Winning in a World Transformed by Social Technologies*. Boston: Harvard Business School Press, 2008.

McLuhan, Marshall. *Understanding Media: The Extensions of Man*. Cambridge, MA: The MIT Press, 1994.

Mezrich, Ben. *Accidental Billionaires: The Founding of Facebook, a Tale of Sex, Money, Genius and Betrayal*. New York: Doubleday, 2009.

Mulholland, A., N. Earle. *Mesh Collaboration: Creating New Business Value in the Network of Everything*. New York: Evolved Technologist Press, 2008.

Palfrey, J., U. Gasser. *Born Digital: Understanding the First Generation of Digital Natives*. New York: Basic Books, 2008.

Rice, Jesse. *The Church of Facebook: How the Hyperconnected Are Redefining Community*. Colorado Springs, CO: David C. Cook, 2009.

Schawbel, Dan. *Me 2.0: Build a Powerful Brand to Achieve Career Success*. New York: Kaplan Publishing, 2009.

Shapiro, C., H. Varian. *Information Rules: A Strategic Guide to the Network Economy*. Boston: Harvard Business School Press, 1999.

Shih, Clara. *The Facebook Era: Tapping Online Social Networks to Build Better Products, Reach New Audiences, and Sell More Stuff*. Boston: Prentice Hall, 2009.

Shirky, Clay. *Here Comes Everybody: The Power of Organizing Without Organizations*. New York: The Penguin Press, 2008.

Solove, Daniel. *The Future of Reputation: Gossip, Rumor and Privacy on the Internet*. New Haven: Yale University Press, 2007.

Solove, Daniel. *Understanding Privacy*. Cambridge, MA: Harvard University Press, 2008.

Sunstein, Cass. *Infotopia: How Many Minds Produce Knowledge*. New York: Oxford University Press, 2006.

Tancer, Bill. *Click: What Millions of People Are Doing Online and Why It Matters*. New York: Hyperion, 2008.

Tapscott, Don. *Grown Up Digital: How the Net Generation Is Changing the World*. New York: McGraw-Hill, 2008.

Turkle, Sherry. *Life on the Screen: Identity in the Age of the Internet*. New York: Touchstone, 1995.

Vander Veer, E. A. *Facebook: The Missing Manual*. Sebastopol, CA: O'Reilly, 2008.

Weber, Larry. *Marketing to the Social Web: How Digital Customer Communities Build Your Business*. Hoboken, NJ: Wiley, 2009.

Winograd, M., M. Hais. *Millenial Makeover: MySpace, YouTube & the Future of American Politics*. New Brunswick, NJ: Rutgers University Press: 2008.

Wright, Robert. *Non-Zero: The Logic of Human Destiny*. New York: Pantheon Books, 2000.

Zuniga, Markos Moulitsas. *Taking On the System: Rules for Radical Change in a Digital Era*. New York: Celebra, 2008.

Index